DINOSAURTOMAN

DINOSAURTOMAN

Mammalian Evolution

Nic Walter Briones

Trafford Publishing
1663 Liberty Drive
Bloomington, IN 47403
1-888-232-4444
Fax: 1-812-355-4082

Order this book online at www.trafford.com
or email orders@trafford.com

Most Trafford titles are also available at major online book retailers.

Please contact the author at nicanor@dinosaurtoman.info
or
P.O. Box 472
Sierra Blanca, TX 79851

Printed in the United States of America.

ISBN: 978-1-4269-4673-8 (sc)

ISBN: 978-1-4269-5208-1 (e)

Trafford rev. 12/02/2010

Trafford PUBLISHING® www.trafford.com

North America & international
toll-free: 1 888 232 4444 (USA & Canada)
phone: 250 383 6864 • fax: 812 355 4082

For my friends EmDalin, Brookes, Shelby, Torri…

Acknowledgements

Thanks to my parents and siblings who supported me to have time to read and collect information, since age two till 38 years old. Thanks also to the following books and references:

Essentials of Geology, 7th Edition
By Frederick K. Lutgens
Edward J. Tarbuck
Laboratory Manual in Physical Geology, 3rd Edition
Richard M. Busch, Editor
Horizons, 6th Edition
By Michael A. Seeds
Geology.com/GeologicTimeScale
Geology.com/Pangea
Funderstanding.com/neuroscience

DINOSAURTOMAN

Contents

Evolution Hypothesis

Human evolution has changed physiognomy and cognitive development of Man. Adaptations to different needs to survive in diversity of environments, such as vast plains, valleys, mountains, and deep dark caves made Man use manual drill and practice for millions of years, and developed brains to invent tools, discovered new ways, and be innovative to the extent of present Earth Sciences and Computer Technology.

In tropical Africa, New Guinea, Australia, isolated and forested for 150 million years, Negroes have lived in only one kind of environment-jungles that have covered the tropical lands for 250 million years. Hence, the Negroes have adapted very little to other surroundings since separating from apes, except loss of body hairs and tails. They retained the simian physiognomies and the cognitive development of life forms with brains that still focus almost entirely on copulating and eating, like any other beasts in the jungles.

In the U.S.A., laws are based primarily on "all men are created equal..." by Abe Lincoln, and "Equality, Fraternity..." by Ben Franklin-men who lacked sufficient knowledge of modern evolution theories, and whose unsubstantiated opinions were mostly influenced by the rabbles of France of the 1770s. These faulty opinions, however, have led to forcible integration by law, of the Negroes into all aspects of Man's life, and promoted interbreeding, due to the Negroes' fixation on copulation especially with inexperienced and vulnerable non-Negro females who are forced in closer contact with the Negroes in schools, work, etc. Just the same, the Negroid offspring has brain development that degenerates backward in evolution by millions of years, and closer to the brains of apes and beasts.

Where are Negro authors, inventors...circa 1700 until today? Soon U.S.A. Man would become inferior to Japanese, Russians, Euros...who have never interbred with these Negroes of Africa, New Guinea, etc.

I don't think that I'm biased, but this paper is a personal hypothesis on evolution of Man and Negroes, based on observations and research of recent scientific findings on origins of life forms on Earth. This is a controversial issue that no one likes to address but needs studying and closer scrutiny in order for Man in the U.S.A. to survive and compete with aggressive and hostile nations on the planet.

Timelines

215 Million Years Ago (MYA)-Age of Dinosaurs. Some large dinosaurs were carnivores and some small reptilians avoided the large meat-eaters by dwelling in deep dark caves and caverns.

96 MYA-Age of Dinosaurs-The small dinosaurs adapted to the dark cave environments by changing skin pigmentations to light and white colors, just as fishes in the dark depths of oceans turned into white fishes, after living in the dark with no sunlight for millions of years. And the small cave dwelling dinosaurs evolved into early cave-dwelling Man.

In forests of Africa, Australia, New Guinea…some dinosaurs adapted to dwelling in the trees, and evolved into gorillas and apes, contemporary with the cave-dwelling Man.

65 MYA-Extinctions of the Large Dinosaurs-The cave-dwelling Man survived the extinctions of the large carnivores, and came out of the dark caverns to begin dominion of the surface of the Earth. Some gorillas and apes also survived and began life on land to evolve into the Negroes, by shedding body hairs and tails, but retained simian physiognomies and brains that are still closer to the gorillas and apes than to Man.

March 2009 A.D.-Age Of Mammals-Man with brains exposed to more than 150 million years of innovations and adaptations to diverse environments in order to survive, developed brains to discover new ways and means of providing for and defending families, and making life easier.

The Negroes lived in only the forests, until the 1700s, hence lack the brains of Man and need to be incubated for 150 million years, in order to catch up to evolutional level of present Man. In my opinion, Mr. Obama is not fit to lead the U.S.A. till 150 million years from now!

Patterns

The following life forms improved upon the preceding ones. Examples: Streamlined fast-moving fish with inner skeletons had improved on the cumbersome outside skeletons of the slow-moving Trilobites.

Geographical locations and environments directed physical and cognitive development and adaptations of the life form. Examples: Smaller dinosaurs lived in caves and caverns for safety against the large carnivores then adapted to life in the darkness by becoming white-skinned Man after 150 million years, just like white fishes found today, living in the ocean depths not penetrated by sunlight. In the forests of isolated Africa, Australia, etc. some dinosaurs adapted to dwelling in trees by becoming gorillas and apes. Only after the extinctions of the large reptiles, 65 million years ago, did the gorillas and apes began life on the surface of the Earth to become the Negroes.

The hands of the bipeds were the keys to the development of the Mammalian and Thinking Brains around the Reptilian Brain. Example: 150 million years of drill and practice, and rote repetitions of groping, grasping and holding stones and sticks developed memory. Innovations in the use of the stones and sticks in hunting, and as defensive tools against the large carnivores followed the rote memory and developed thoughts.

Available food was a major factor and motivator for changes in characteristics of the life forms. Example: The fish developed limbs to be able to travel on land where the food was abundant. Some dinosaurs grew into giants due to the endless supply of food.

Life forms could not reverse evolution to a previous life form but could degenerate in brain development and physiognomy, by interbreeding with the previous life form. Example: Daughters and wives of Man indulging in bestialities with apes, gorillas and simians, etc.

Reptilian Brain

Neuroscience

Neuroscientists have found that the brain of Man consists of Reptilian Brain-for sensory and motor functions; Mammalian Brain-for memory, emotion, and biorhythm; and Thinking Brain-for cognition, language, reasoning, and higher intelligence. The brains and nervous system have control of learning processes.

Man can achieve continual learning and development of intellect throughout a lifetime. Mental exercises alter the brain structure for better usage, next time. I agree with these research findings.

INSIGHTS: Where could Man's reptilian brain originate?

Geologists have found that four Biological Ages had passed Earth's History, with an average 119 million years per Age (excluding the ongoing Age Of Mammals). Started by invertebrate Trilobites, some trilobites developed inner skeletons to become faster-moving Fishes. Some fishes liked living on land as well as in water, and became four-legged Amphibians. Some amphibians preferred use of two limbs and arms to become biped Dinosaurs with two hands and developed Reptilian Brain.

The small dinosaurs avoided becoming food for large meat-eaters, by dwelling in dark deep caves for 150 million years-Enough time to develop the Mammalian and Thinking Brains around the Reptilian Brain to become cave-dwelling Man!

In forests of Africa, New Guinea, Australia, isolated for 220 million years, some dinosaurs adapted to life in trees by becoming tree-dwelling gorillas and apes. When the large reptiles were wiped out 65 MYA, the cave-dwelling Man survived the extinctions and came out of the caves to begin dominion of the planet. Some gorillas and apes also survived and began life on land. A few MYA, the gorillas and apes became Negroes. Hence the evolutional heritage of Man and Negroes traces back to the Age of the Dinosaurs, all the way to the Age of the Trilobites!

Where else could Man's Reptilian Brain come from?

Field Trip: Go to Fish-Seafood stores and find out which are Man's remotest biologic relatives. Explain in class reasons of your choice. (Answers: Lobsters, shrimps etc. far cousins of the exoskeleton trilobites).

Evolutional Lag

The exoskeleton Trilobites lived in Water. After 119 million years of slow mobility, some of the trilobites became endoskeleton for faster-mobility in the waters and evolved into fishes. After 119 million years, some fishes became four-legged amphibians to be able to survive in the waters as well as on land, for more food and safety.

After 119 million years some amphibians became bipeds for better survival on the land and developed reptilian brains to become Dinosaurs that could live on the land, in the waters, and in air.

Some dinosaurs became cave dwellers in Laurasia, Americas and Antarctica to avoid the large carnivores aboveground. After 119 million years, the cave-dwelling dinosaurs became Caveman with Mammalian and Thinking Brains, and with light-colored skins due to 150 million years of life in the dark caves not penetrated by sunlight. In forests of Africa, New Guinea, Australia, etc. some dinosaurs adapted to dwelling in the trees by becoming apes and gorillas with no Thinking Brain, contemporary with the Caveman.

Then the Extinctions of the large reptiles and carnivores occurred on the surface of Earth, 65 million years ago. The Caveman survived in the caverns and came out of the caves to begin dominion of the Earth as dominant species on the land. Some apes and gorillas also survived and began life on the land. And some of the apes and gorillas became bipeds again. The apes and gorillas began developing mammalian and thinking brain to emerge as Negroes.

By 2010 A.D. the Caveman has developed a civilization consisting of advanced Sciences, Computers and Technology to propel Man into the Space Age. Some apes and gorillas also managed to shed body hairs and tails to become the Negroes, with unknown parts apes and gorillas in the Negroes' systems.
FUTURE MAN-54 Million A.D.-Future Civilization of future Man.

Negroes will become the equivalent of the Caveman with Zero-Part Simian in their systems. And the Negroes' evolutional lag behind Man equals 31+65+54 or 150 Million Years!

Bipeds And Giants
What Develops Thinking Brain?

Two-footed life forms have more chances of developing the thinking brain than four-footed ones. With the other limbs used as arms with hands to grasp and hold food, sticks, stones, and objects, and beginning with accidental usage of the sticks, stones, and objects into tools or weaponries could result to memory via long-term repetitive drill and practice for millions of years! Creative innovations and imaginative inventions could eventually follow.

Ground-crawlers such as four-footed lions, cats, dogs, etc. would take a longer time to develop the thinking brain than squirrels, rats, mice, etc. that have already learned to move around on two feet while grasping food and objects with their shorter limbs or hands. Gorillas, apes, and monkeys that dwell in trees would also have less chance of developing the thinking brain, because of usage of the hands to only grab branches with no need for creative use of the branches as tools or weapons.

The biped dinosaurs that had dominated the Earth for 150 million years had developed the Thinking Brain and Mammalian Brain, AFTER 119 MILLION YEARS of evolution. The biped reptilians became GIANT MAN with Man's triad Brains, long before the extinction of the large dinosaurs 65 million years ago. That the GIANT MAN existed and lived above-ground while the large dinosaurs still roamed the surface of the planet, is indicated by giant man's fossils found all over America, India, Eurasia...and by the fossilized footprints of the Giant Man striding side-by-side with fossilized tracks of the large biped dinosaurs, found in Texas...also giant man's remains found in California with multiple upper rows of teeth, giant man's remains found in India. There was also a golden chain that was found imbedded in coal somewhere in Pennsylvania, U.S.A.

Hence, another type of the Dinosaurs besides the CAVE-DWELLERS and the TREE-DWELLERS, were the SURFACE-DWELLERS. The surface-dwelling dinosaurs such as the BIPED Tyrannosaurus Rex had become the GIANT MAN after 119 million years of evolution, complete with fully developed Mammalian

and Thinking Brains, 31 million years before the Extinctions of the large Reptiles.

Most of these Giant Man perished during and after the Extinctions 65 million years ago, but there were most likely survivors that continued living with Man, as mentioned in myths and ancient legends, such as the Cyclops of Homer...Artifacts left by the Giant Man are Stonehenge of England, Sculptures on Easter Island, Ponape's cyclopean wall-like structures in Micronesia, Building blocks at Machu Picchu in the Andes.

Giant Man fossils could be discovered in 65 to 96 million-year-old sedimentary rocks. The rocks on which the fossilized footprints and tracks were found, as well as the coal imbedding the golden chain, would also be between 65 and 96 million years old.

Life-Forms Parade

Trilobite	Fish	Amphibian	Dinosaur	Mammal			
Lobsters	Carps	Frogs	T. Rex	Giant Man	Man	Apes	Negroes
Invertebrate	Vertebrate						
Aquatic	Aquatic	Land- Water	Land- Water- Air	Land	Land	Trees	Land
Clumsy	Streamlined	4-Footed	Biped	Biped	Biped	Four- Limbed	Biped
Slow Mobility	Fast Swimmer	Fast Mobility	Fast	Fast	Fast	Agile	Agile
			2 Hands	2 Hands	2 Hands	4 Limbs	2 Hands
			Reptilian Brain				
				Mammalian (Mmmln) Brain	Mmmln Brain	Simian	Mmmln Brain (X-Part Simian)
				Thinking Brain	Thinking Brain	Simian	Thnking Brain (X-Part Simian)

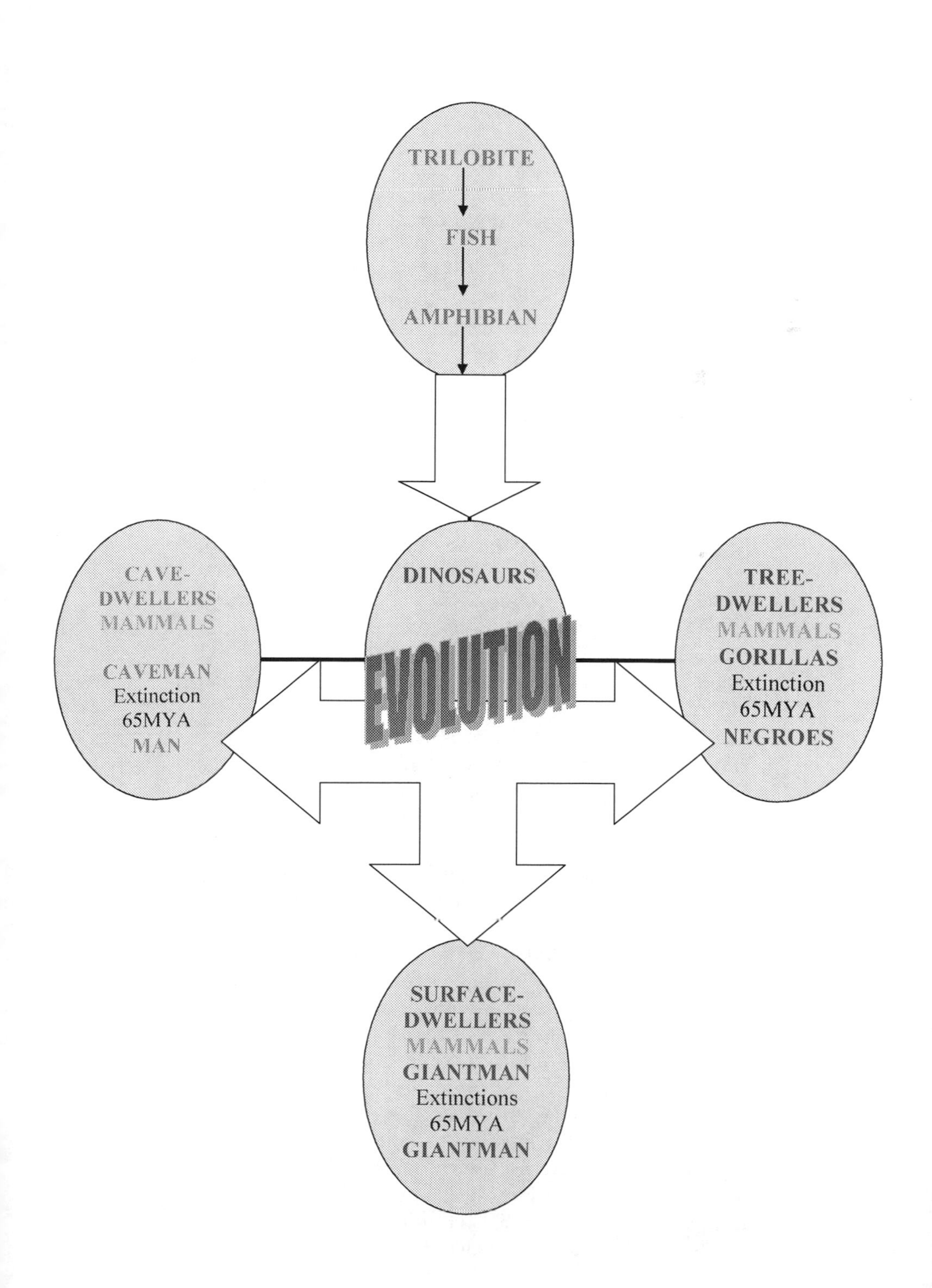
TRILOBITE
FISH
AMPHIBIAN
DINOSAURS
EVOLUTION
CAVE-
DWELLERS
MAMMALS
CAVEMAN
Extinction
65MYA
MAN
TREE-
DWELLERS
MAMMALS
GORILLAS
Extinction
65MYA
NEGROES
SURFACE-
DWELLERS
MAMMALS
GIANTMAN
Extinctions
65MYA
GIANTMAN

Pangea Inclusion

The formation and break-up of super-continent Pangea was considered as irrelevant in Evolution of Mammals. It was established that:

1. Mammalian Man originated from African gorillas and apes.
2. Man and African Negroes have a common origin-the Tree-dwelling African apes and gorillas.
3. Changing environments resulting from the Break-Up of Pangea, have nothing to do with the evolution of the Mammalian Man and the Negroes!
4. Missing-link fossils are just waiting to be dug-up somewhere in Africa.

The formation and the break-up of the super-continent Pangea, about 250 million years ago, was a primary factor in the evolution of mammalian Man and the African Negroes from the reptilian Dinosaurs! The Dinosaurs that had dominated the Earth as dominant Life Form for 150 million years, had divided into different types, due to the changing ENVIRONMENTS where the life form lived, that resulted from the break up of Pangea..

One type of the reptilians consisted of Cave-Dwellers-smaller dinosaurs that avoided the large reptilian giants living on the ground, and forced to live in the caves with dark environments without sunlight, for 150 million years. In the process, the cave-dwelling dinosaurs adapted to the dark surroundings in the deep caverns, and changed skin pigmentations into pale white colorings.

After 119 million years of life in the caves, the cave-dwelling biped dinosaurs evolved into mammalian caveman, complete with Mammalian and Thinking Brains developed around the Reptilian Brain, and formed via the use of HANDS through millions of years of drills and practice, grasping and holding sticks and stones, etc. and using the objects as survival tools and weapons against the large meat-eating predators. Creative improvisations, innovations and imaginations followed imprinted memories of defending families that further formed feelings and emotions of love, fear, anger, etc. to develop the Mammalian Brain.

The Caveman appeared in the Americas, Europe and Laurasia-continental portions of Pangea that were interconnected by land bridges. The Caveman had developed an unknown civilization during the remaining 31 million years of the Age of the Reptilians on Earth, up to the Extinctions of the large dinosaurs, 65 million years ago.

In the isolated continental parts of Pangea-Africa, New Guinea, Australia...that drifted alone as island-continents-another type of the dinosaurs had adapted living in the trees of the vast tropical jungles, to become TREE-DWELLING DINOSAURS! The biped dinosaurs had reversed back to four-footed beasts and back to non-use of the hands except to grasp vines and tree branches to swing from tree to tree-resulted to shrinkage of the Thinking Brain and dormant creativity and imaginations. Only the simian limbs with arms and hands grew as the longer limbs.

After about 119 million years, the tree-dwelling biped dinosaurs degenerated into four-limbed gorillas and apes with black hides due to lifetime after lifetime of exposure to sunlight. For 31 million years more, the gorillas and apes became agile four-limbed swingers in the jungles of Africa, New Guinea, Australia...but did not find a need to redevelop the Mammalian and Thinking Brains, until the Extinctions of the giant reptilians, 65 million years ago.

The jungles had disappeared for years and deprived some surviving apes and gorillas, the tree habitats. Forced to live on the ground, some gorillas and apes became bipeds, and restarted to find use of the hands to grasp and hold food, stones, and sticks...as tools, weapons, and the Thinking Brain commenced developing. Close-knit families of some biped gorillas and apes emerged, and sometime between 65 million years ago and 2010 A.D. some biped gorillas and apes succeeded to shed body hairs and tails to emerge as the Negroes!

Due to an average of approximately 119 million years in the evolutional time frame, is necessary for one life form to improve into the new life form, i.e. from the fishes to frogs, or from the gorillas and apes to the Negroes, the Negroes contain unknown parts (xx-parts) gorillas and apes in physiognomy and brain development as of today. It follows that 54 million years more, in the evolutional time frame, are necessary for the Negroes to acquire ZERO-PART (00-part) gorillas and apes in their systems. That time will be in year 54

Million A.D. And the Negroes will become equivalent to the Caveman in the evolutional time frame.
The Caveman became Man of 2010 A.D. with all the adaptations to survive in ever-changing environments-wars, famines, pestilence-driving forces that gave birth to creativity, for Man to have Earth-Space Sciences, Advanced Technology and Computers…to dominate the planet as the dominant species.

The Pangea factor could NOT be excluded in analyzing the Evolution of Mammalian Man and Negroes. And Pangea is a fact as shown by the EXACT FIT of the Atlantic edges of Africa and South America-indicating that the two continents were once upon a time, coupled together as one large continent, and Africa drifted alone in the Atlantic Ocean as an ISOLATED CONTINENT for 250 MILLION YEARS!

This is the unknown factor that scientists and Man have completely ignored in the search for the Origin of Man and the African Negroes.

My personal project is to search for the fossil-link between the Caveman and the cave-dwelling dinosaurs in rock strata between 100 and 250 million years old, in the Americas, Eurasia, and Antarctica. In the same strata, I'll also search for the fossil-link between the tree-dwelling dinosaurs and the gorillas and apes, in Africa, New Guinea, Australia, etc. And I'll search for the fossils linking the Negroes to the gorillas and apes, in 65 million-year-old (maximum) sedimentary rocks in Africa, New Guinea, Australia, etc.

Catfish To Frogs

Amphibian Evolution

Thinkology: There are many ways to integrate critical thinking in classroom activities. One way is to give students, means to understand meanings in texts by using animation, so the students could make better decisions about issues in question.

For example: the issue is origins of Amphibians. How do the frogs come about? Using slides and videos, a teacher could show an animation of a catfish. The catfish crawls out of the water of a bank of a river, and propels itself on the land by its side fins.
Then, gradually the fins transform into feet, after millions of years. An audio device could explain that the feet-forming process is due to adaptation by the catfish in response to its need to travel on the land, for more food and safety besides the food supply in the waters. A student might ask questions: Why would the catfish like to live on the land? Why does the catfish take millions of years to make feet?

The teacher could explain that catfish brains focus entirely, every moment of its life, on eating and not being eaten, hence the catfish has no time to leisurely grow limbs in a hurry…or it liked the land better to avoid being made into "fish sandwich" by bigger fish. After clearing up and answering the students' questions about the evolution issues of the frogs from the catfish, the teacher could help the students make decisions about the issue.

Kidspiration: In small groups, students use the Internet to find information about ongoing loss of Florida plants and trees, to commercial developments, and predict possible effects on Florida after 50 years. Then the students are to classify their findings according to least devastating to worst scenarios they could imagine. And what the students could recommend as solutions.

The possible findings could be:

1. Massive out-of-control erosion of topsoil so the trees or plants could not take root and grow, then, liquefaction, and sinkholes.
2. Protect Florida from floods by allowing only more than 45-story buildings for housings, subdivisions, churches, commercial malls, schools, and government offices.

3. **Immediate stop in production, sale, and use of Internal Combustion Engines hence eliminating pollutions due to unburned fuels, roads and highways, parking lots, etc.**
4. **Use alternatives such as battery-powered Solar Dirigibles for transportation and cargo-transports. Finally, the students could sum up their findings for presenting to the whole class, using audio-visual aids and presentation graphics.**

Gorilla Migration

How The Earth Was Won

According to current findings by a team of 20 scientists who are leaders in geology, paleontology, Neurobiology, Math, Physics, etc. and who are recognized experts in their respective fields by the schools and universities that employed the experts, Man's origins were African apes, gorillas, and baboons with simian physiognomies and zero beast-brains that focus entirely on feed and copulate, circa a few million years before the Jewish virgin started to give birth to children, sired by an unknown father.

Then the gorillas, apes, and baboons crawled on four feet, swung from tree to tree, across the vast jungles and built rafts and ships to cross oceans and seas, while the super-continent Pangea, started to break up and Africa drifted away from South America, to populate Earth in a massive gorilla migration from Africa! In route, the gorillas, apes, and baboons began to change in physiognomies-black skins to lighter hides, simian body hairs and eyes changed to blonde, auburn, blue, green, etc., gorilla lips became lighter ruby-red lips…and that by the time the gorillas and apes reached Japan, the gorillas and apes became slit-eyed, bucktoothed, bowlegged sneaky Japs with thick yellowish hides!

How come the Negroes of Africa still retain simian features is deliberately overlooked for fear of losing their jobs…are beyond me. I guess the answer is the perverse obsessions of Man-sacrifice all-daughters, wives, family members…except personal savings, to become a senator, governor, etc.

That the present Man and the Negroes originated from the dinosaurs could be traced from the Reptilian Brains that these mammals retained. The Reptilian Brain could only come from the Dinosaurs! The professional opinions and findings by the scientists who cooperatively studied every aspect of remains of a hominid that was recently dug up in sedimentary rocks in Africa, indicate that:

1. **The hominid fossil is the origin of Man and Negroes.**
2. **The hominid is a look-alike of the present Negroes except for hairs covering its hide, as depicted through computer enhancements.**

3. The Negroes were still look-alikes of the gorillas and apes circa 5 million years ago.

4. The Negroes separated from the gorillas and apes, as recently as a few million years ago or less, but less than 5 million years ago!

I agree with the team's findings that were published by Seed Magazine about 2007. The professional opinions clearly support NWB's Evolution Hypothesis, that:

1. The Negroes separated from the gorillas and apes as recently as a few million years ago, or less.

2. The Negroes are still more gorillas and apes than Man, in physiognomies and brain developmental level that focus almost entirely on copulate and eat, like any other beasts.

3. Chairman Ho is still not fit to lead Man and the free world, due to an evolutional lag of 150 million years, in the Chairman's Thinking and Mammalian Brains. Even if the Chairman is half Man half Beast now, 75 million years are still necessary for the Chairman to at least become, Caveman HO in the evolutional time frame.

Story Of Man

Why Man Is Different

215 Million Years Ago

Super-continent Pangea began to break up. Africa, New Guinea, Australia, India...drifted on the oceans as isolated islands. Dinosaurs were dominant on Earth. Some dinosaurs in Laurasia, Americas...avoided the large carnivores by living underground, in deep caves and caverns not penetrated by sunlight, to become Cave-Dwelling dinosaurs. Some dinosaurs in tropical jungles of Africa, Australia...lived in the jungles to become Tree-Dwelling dinosaurs.

96 Million Years Ago

After 119 million years, the cave-dwelling dinosaurs had adapted to the dark environments of the caves, by changing skins into lighter white colorings due to lack of sunlight, and became the Caveman! The tree-dwelling dinosaurs in Africa, Australia...had adapted to the sunlit environments to become four-limbed gorillas and apes with thick black hides.

Some large biped dinosaurs on the surface of the Earth, also changed into Giant Man...complete with mammalian and thinking brains, contemporary with the Caveman and the gorillas and apes. The Giant Man had lived in Laurasia, in the Americas, in India...

31 Million Years Later

The Caveman had developed an unknown civilization in the caverns, and the Giant Man had also developed a civilization that is unknown to present Man. The gorillas and apes multiplied in the African jungles...

65 Million Years Ago

The Giant Man and the giant dinosaurs were wiped out in the Extinctions. The Caveman and some Giant Man survived the extinctions, and ascended aboveground, and began dominion of the planet as Man. Another ancient civilization of Man was begun. Some gorillas and apes in Africa...survived the extinctions and began to live on the land, not in the trees. Some gorillas again became bipeds.

2009 A.D.

Albert Einstein, Nic Tesla, Newton, Gates...propelled the current civilization of Man, including Advanced Technology, the Sciences, and Computers. The gorillas and apes became Negroes with unknown (x-part) simian in the Negroes' systems.

54 Million A.D.

Future civilization of future Man will take place on the Earth…The Negroes will at last, have 00-part simian in the Negroes' systems, and will become the equivalent of the Caveman, in the Evolutional Time Frame.

Story of Life Forms

Geologic Time

540 million years ago, slow-moving trilobites with skeletons crudely placed outside their bodies, became dominant life form in oceans of Earth. Then an improved design of sea species called fish, with the skeleton inside the body replaced the trilobites as dominant in the oceans. The fish were streamlined, faster-moving than the trilobites.

On land where plants, trees, and animals began to thrive, amphibians appeared with four limbs but have closer resemblances to the fishes, especially the catfish. Then 215 million years ago, four-footed and biped dinosaurs emerged as the dominant species on the planet. The biped dinosaurs have two arms with two hands and two limbs instead of the four-limbed amphibians, which mean that dominant bipeds were superior to the four-legged beasts in terms of survival. It follows that the superiority in terms of survival, by the bipeds could be due to the arms and hands that the bipeds possessed. Furthermore, the bipeds' hands had initially produced Reptilian Brains on the biped Dinosaurs. Naturally, the dumb and undoubtedly stupid four-footed plant-eating dinosaurs became food for the biped carnivores.

The biped dinosaurs had divided into a number of types: Surface-dwellers, Cave-dwellers, Tree-dwellers, Land and Air dwellers, Water-dwellers…The Cave-dwelling biped dinosaurs, the Surface-dwelling biped dinosaurs, and the Tree-dwelling four-limbed dinosaurs are the types of interest in this paper, because of their eventual EVOLUTION into Man and Negroes.

The Cave-dwelling biped dinosaurs became the biped caveman. The Tree-dwelling four-limbed dinosaurs became apes, chimpanzees, baboons, and gorillas. The Surface-dwelling biped dinosaurs became the Giant Man…after an average 119 million years of living and adapting to different environments on the surface of Earth, resulting from the Break-Up of the super-continent Pangea, about 215 million years ago. It so happened that the 150 million years of the Age of the Dinosaurs occurred during the Pangea break-up. And Pangea is a fact, as seen in the exact fit of the Atlantic edges of the continents of Africa and South America, indicating without question, that these two continents were coupled together as one continent, and drifted apart from each other, alone as an isolated Africa, for about 200 million years!

The Caveman with thinking brain and mammalian brain had lived in the caves for 31 million years, until the extinctions of 65 million years ago. Then the caveman came out to become the dominant species on the surface of the Earth, to the extent of present Technology, Sciences, and Computers.

The Tree-dwelling dinosaurs became gorillas, apes, baboons, and chimpanzees, after 119 million years of life, swinging in the trees and jungles of Africa…For 31 million years more, the apes and gorillas became agile swingers in the jungles but had not developed any civilization of any kind in the jungles-no houses, no clothes…in other words, they lived as beasts with zero beast brains, while the Caveman and the Giant Man were already using fire to cook meat, creating and using tools and weaponries, etc. Only after the extinctions of 65 million years ago, did the surviving apes, gorillas, and chimps, began life on the lands of Africa. Sometime between 65 million years ago and 2010 A.D. the gorillas and apes became Negroes with unknown (x-part) part simian in their mammalian brain and thinking brain. In other words, the Negroes of today have 54-million-year worth of evolution to go, in order to reach the evolutional level of 2010-A.D. Man, based on the average 119 million years required for a life form to improve into the next life form, i.e. from dinosaur to caveman, or from dinosaur to the gorillas. And as the numbers are close approximations, based on the Geologic Time Scale (Chart found in any Geology textbook), and the fact of Pangea…then Man (the daughters and wives of Man) in the U.S.A. and Great Britain…who have been copulating with the Negroes since the late 1800s, are definitely meshed up by bestiality!

Hence, the mammalian Man and Negroes, descended from the dinosaurs as proved by the Reptilian Brain that these mammals retained. Man's remotest known relatives are lobsters-far cousins of the exoskeleton trilobites, and Man's close biologic relatives were the cave-dwelling biped dinosaurs and the caveman. The Negroes' close cousins were the four-limbed tree-dwelling dinosaurs, and the gorillas and apes of Africa, New Guinea and Australia..

The Mammals are still in ongoing processes of evolution due to adaptations to ever-changing environments, food supplies, ways of living, etc.

10% Functional Brain

Man is known to use only ten percent of three brains. Which parts of the triple brains of Man, are used? With the use of only ten percent of Man's brain, even beasts with zero-percent thinking brain could compatibly live, side-by-side with Man. The thinking brain of Man is the key to Man's creative, imaginative, and innovative superiority to survive in wide variety of environments.

A Sample Chart:

Life Form	%Reptilian	%Mammalian	%Thinking	%Simian...	%Total
BClinton	5.0000	4.9995	0.0005	00.0000	10.0000
JFK	5.5000	4.0000	0.5000	00.0000	10.0000
ALincoln	6.0000	3.5000	0.5000	-2.0000	08.0000
RReagan	3.0000	6.9100	0.0900	-1.0000	09.0000
GBush	4.5000	5.0000	0.5000	00.0000	10.0000
GWBush	5.0000	5.0005	0.9995	00.0000	11.0000
BObama	3.0000	5.9999	0.0001	-50.0000	-41.0000
AEinstein	1.0000	0.5000	11.0000	00.0000	12.5000
NWB	1.0000	1.0000	8.5000	00.0000	10.5000
Babyman	5.0000	7.9999	3.0001	00.0000	16.0000
Beasts	7.0000	8.0000	0.0000	00.0000	15.0000
Gorillas	8.9999	0.0001	0.0000	-99.0000	-90.0000
Parrots	8.9999	0.0000	0.0001	-10.0000	-01.0000
Mynahs	8.9999	0.0000	0.0001	-11.0000	-02.0000

Animal Rights

Make A Guess

Man has never tolerated cruelty to animals. Beat a dog or a cat, and you'll be liable for fines or criminal misbehaviors of some kinds. Humane Societies made sure that the animals are given shelter and enough care to live a normal life.

In the U.S.A. there are animals that are literally tortured all days of their existence, yet the Humane Societies averted their eyes, deliberately from seeing the mistreatment of these animals. Firstly, these poor animals (I'll call them "they") are deprived of sunlight till the last day of their lives on Earth. They are sheltered in darkened concentration camps, completely isolated from the rest of the animal world.

Then they are force-fed with enhanced diets in order to hasten body growth. Once their bodies become large enough, they are again force-fed with reproductive hormones, to force them to reproduce. Because, like most beasts, they reproduce at night, the darkened environment in which they are kept, altered their senses of day or night (to them, it is always night time!), hence they abnormally reproduce several times in a twenty-four-hour day. Besides losing their rights to normal growth into adulthood, they are forced to reproduce like the adults while still babies.

Their being still very young then forced to reproduce several times a day, the forced feedings with enhanced diets that also include high dozes of antibiotics, and the isolations from other animals in the free world…reduce their life spans to only nine weeks, compared to normal life span of their kind of animals living free, of up to 3 years.

They are also deprived the company of their male kinds, hence, they are forced to live like nuns in convents. How they are able to reproduce is beyond my comprehension.

The last day of their earthly existence, they are shipped in multi-storied cars to death-camps; at least they are not death-marched like the Asian natives and POWs driven by the Slit Eyes, to Mukden. Some of them arrive DOAs at the camps.

Then their heads are chopped off. While still kicking, the headless bodies are passed through hot showers and scalded for easier cleansing of their skins;

then they are disemboweled. Those bodies that are underweight or overweight go leftward-their flesh are torn from bones and conveyed to grinders-skins, fats, cartilages, and all. The chosen ones go straight to the ovens!

They are Man's close biologic relatives-even closer than Negroes due to lack of transitional links because of special environments, such as the African gorillas and apes in isolated Africa for the Negroes. Like Man, they descended directly from the biped dinosaurs after evolution of 119 million years! They did not succeed to develop thinking brain due to the size of their heads.

The bucktoothed Slit Eyes eat them raw, after flaying them alive…Chomp, slurp, chomp-I guess these bowlegged bloodsuckers live mostly to bloody centenarians, because of devouring raw live meat and sucking blood straight from the neck!

Martians And Man

Why does Man like to live on Mars?
How shall Man live on Mars?
How will the Martians respond to the arrival of uninvited Man?

It is always fun to explore new lands and to possess the lands irrespective of native inhabitants. Either by force through swords and guns, or by deceptions, in the names of gods so the natives could die by the sword or guns, Christianized...

One motivation of Man going to Mars is to exploit Martian resources. Would the native Martians give in and allow Man to just take over the Martian lands? I guess there would be massacres of Man, especially if the Martians are as advanced in Science and Technology as Man knew the Martians are. Why pretend not to know?

How Man would live on Mars, according to current multimedia information is to work the Martian surface by terraforming. This is a fallacious gimmick. Firstly, transforming the Martian surface into livable habitat for Man, could take thousands or millions of years. And enormous funds in US $Trillions would result to hollering about wastes due to prioritizing.

It is common knowledge in the 2000s that the Martians exist inside the bowels of Mars! Who were these Martians? Most likely, they are the former gods of pre-historic Mesopotamia in the Middle East, who left Earth after transporting gold from the Earth to their home planet, for about a million Earth years!

In consequence, the planet Earth was transformed into a midget planet, by weight, from a loaded, super-jumbo 747 to an ultra-light glider in terms of weight, which could be a major cause of inclement weathers on the planet Earth today! So Man has to first, fight the ultra-high tech Martians in order to dislodge and dispossess them of their subterranean abodes and lands. It would be easier if the Martians are like the Indians (the Hollywood type, with hook noses, red skins, riding bareback on wild mustangs...not the Bombay type); have the same lifestyles as Man...you know-devouring processed meat and munching baked bleached flours with refined sugars, etc. and lying most of the time in order to please crowds and remain in

salaried post, hammering mostly good-for-nothing laws…

But if the Martians are living like ultra-high tech bees, with ultra-high tech defensive and offensive weaponries-inconceivable to present Man-in their impregnable beehives…it would be better for Man to stay put and solve the problems that face Earthlings, now!

Change Man's life styles! Only then could Man live side-by-side inside Mars with the Martians. And be ready to massacre the Martians, by including the sneaky Slit Eyes among the first Martian arrivals!

Why? Why? Why?

Why can't some members of Man accept the evolution of Man as separate from Negro evolution? **Because of self-interests of personal gain, such as financial gain; and because of beliefs in dead-end creators that promote that a mammalian god created the mammalian Man! The belief also promotes rampant indulgences of ignorant daughters and wives of Man to copulate with Negroes, hence reproducing Negroid offspring that are more gorillas and apes than Man, and have now overwhelmed the U.S.A. in numbers as majority inhabitants, as proved by BObama in the Oval Office.**

Another belief is that Man and the Negroes originated from common ancestors-the gorillas and apes of Africa-thus disregarding environmental effects of Pangea on evolution of Man; also ignored closer resemblances in physiognomy and agility of the Negroes to the gorillas and beasts than to Man. These beliefs are taught in education of Man's children, by politicians who gave more emphasis on sports, wherein the beastlike agility of the Negroes is mistaken for intelligence!

Why can't all members of Man believe that Mammals descended from the once dominant reptilian Dinosaurs? **The answers lay on what are considered sacrosanct laws that were made by ignorant politicians before the advent of Earth-Space Science and before the advancements in the Sciences, Technology, and Computers. Again, the beliefs in many kinds of gods and creators by many cults and religions, have directed Man's attention to blindly follow the priests, preachers, popes, imams, etc. to continue believing in false unsubstantiated opinions of clergies, etc.**

Why are the daughters and wives of Man the ones who are more vulnerable to bestialities with the Negroes than the sons of Man? **Because by law, everything in life styles in the U.S.A. bring non-Negro women in closer contact with the Negroes from childhood to adulthood through schools, work, sports, and U.S.A. society in general…and the Negroes beastlike survival instincts that focus entirely on copulate, copulate, eat, eat, copulate…victimized the** guileless, innocent, young daughters and wives of Man, sooner or later. The male members of Man have varied interests and creative pursuits of honor, achievements in the sciences, etc. occupying Man's thinking brain, and find the still ape-like female Negroes unattractive, ugly, and unsuitable to become mothers of Man's children, as mates.

Why do some educated members of Man support the unsubstantiated opinion that Man and Negroes have common ancestors-the apes and gorillas of Africa? One main reason is economical interest to exploit African resources, and to please the Negroes, who are the native inhabitants of the African lands, thus have first possessions of the materials in contention. Gold, Uranium, diamonds, oils…are among the resources coveted by the educated members of Man. Why not just exterminate these mammals that have no use of these nonrenewable materials, is more decent than selling these beasts arms and weapons to kill each other, in my opinion. Another reason is some educated members of Man, are ignorant of the recent scientific findings on the evolution of life forms on Earth. Despite the scientifically proven facts of the Geologic Time Scale, the Formation and Break-Up of the super-continent Pangea, and the Triad Brains of Man as found by neuroscientists, these self-interested motivations of the educated members of Man still advocate the idea of common ancestors by Man and Negroes from the tree-dwelling African apes. Why not from the tree-dwelling apes, gorillas, and chimpanzees of Australia, New Guinea, Borneo, India…? These are also island-continents, like Africa that drifted alone on the oceans for 220 million years during the Break-Up of Pangea.

And despite the close resemblances of physical features of the African Negroes to the African gorillas and apes, as compared to the diametrically opposite physiognomies of Man, as well as the still beastlike agility of the Negroes in commercialized sports in the U.S.A. the educated Man remains complacent in accepting common simian origins due to laziness in use of Thinking Brain. Or are their thinking brains so washed up, self-mesmerized and hardened by lifetime of religious and political indoctrinations?

Avionics

Slit Eyes

One lesson learned in wartime avionics is that building an empire doesn't need avionics systems or uniformed military. The sneaky slit-eyed Emperor of Japan has used the United States military as "volunteers" and US taxpayers as paymasters, to continue plundering and murdering masses of Chinaman, Koreans, Myanmars, Filipinos…Tibetans since 1939 ongoing to 2010 A.D. The rapes and murders perpetrated on millions of the daughters and wives of Man in Nanking City, China by the Emperor's hordes, then replacements of the dead natives with sneaky Jap settlers, repeated millions of times all over Asia, resulted to present-day Chinaman, Koreans, Myanmars, Mongolians, Taiwanese, Filipinos, Tibetans…as sneaky Japanese in-disguise as Asians, completely fooling the world with stolen native identities, names and dialects or languages. Another result is while all nations on Earth are having population explosions; only Japan has a decreasing population. Where did all those missing Japanese sneak away? Find them in China disguised as Chinaman, in Korea disguised as Koreans, in the Philippines disguised as Dringo Catholics, in Tibet disguised as Tibetan monks (Chinaman disguised as Tibetan monks, and sneaky Slit Eyes disguised as Chinaman).

Generations of the Asian survivors of the Asian Holocaust were doped, for decades, into zombies' slaves with heroine and opium, thanks to Dr. Ishii's Mukden experiments on Live Chinaman and POWs. From Pearl Harbor, Bataan, to Gen. McArthur's "Bomb China" (Bomb the Slit-Eyes military intact in China), to the "Reconstruct Japan" programs that top coated the US politician's hides as Mother Teresa, the intact Japanese military all over Asia, has acquired unlimited licenses to rape, to rob, to mass-murder, to dope whole families of Asians for generations, into slaves since 1939 ongoing to this day!

Dringo spies stole Top Military Secrets for their emperor who developed overnight, Space Avionics Systems in China and Japan, and eventually won the Pacific Wars with the support of the U.S.A.'s "Rearm Japan" to defend the sneaky emperor as the U.S.A.'s staunchest ally, against the Emperor's orchestrated attack by the very bad Koreans (sneaky Slit Eyes disguised as

Koreans!); And the endless transport of Asian loots to Tokyo, along with the US AID of $Trillions to the emperor and to the Asians (the Japs Marcos, Laurels, Aquinos, Estrada...disguised as Dringos), drained the U.S.A. resources, making the Gringos subservient to the Slit Eyes, until 2110 A.D.!

How can this be? The U.S.A. has won all the wars according to Newspapers and TV commercials! **Let us trace the events that had occurred before and after the wars with the Slit Eyes!**

1. **Sneak attack at Pearl Harbor: 300 Slit-Eyes Zeros sneaked down the harbor and made sitting ducks of the US avionics and naval armaments, along with 165000 US-patriot casualties within 30 minutes! Plus the Pearl-Harbor boys and girls, who survived the sneaky attack, have suffered nightly nightmares all the days of their lives! And the Slit Eyes sneaked into the Harbor while Emperor Chop and the US politicians were wrangling in the Oval Office over how to phrase a civilized, proper Declaration of War along with the usual holiday parades and blaring brass bands!**
2. **Bataan Has Fallen: The U.S.A. military in the Philippines and the Dringos also became sitting ducks! The thousands of survivors were finally given time to display themselves in a different kind of parade, called Death March over 400 miles under fiery oriental sun, monsoon rains, and mosquito bites without food, water, and rest. Every Slit Eyes had fun, except the Gringo and Dringo marchers.**
3. **Rape and murder of a million daughters and wives of Man in Nanking City, China: The genocidal Slit Eyes had continually raped the defenseless women, day after day, week after week, for months...murdering children and old women, too young or too old to copulate then doped the survivors to lifetime of enslavement as prostitutes and slave-laborers for the Jap military in China, Koreas, and in Hiroshima, Nagasaki, Tokyo...Where were the male Chinaman? They ran to the mountains leaving their women behind, but were later rounded up and forced to don Jap military uniforms, then shipped to their doom, to the Pacific islands with nowhere to run or hide, to welcome the Gringos, led by Gen. Mc Arthur, in Rabaul, Iwo Jima, Guadalcanal...Hiroshima, Nagasaki.**
4. **Koreas, Mongolia, Myanmar, Vietnam, Indonesia, Philippines: the same rapes, murders, looting of local treasuries and the natives, stealing of the victims**

identities, names, and dialects continued unabated, since 1939, so that all the present-day Asians are now the 10th generation of Slit Eyes disguised as Chinaman, Koreans, Dringo Catholics...

5. General McArthur's Arrival in Japan: The first question that came to the general's head was, "Where are all the sneaky Slit Eyes?" All Gen. McA saw were slit eyed dogs, cats...that made the general very suspicious. The general knew that the sneaky Slit Eyes military were intact in China, Koreas...disguised as mandarins, so Gen. McA cabled Chairman HO to Bomb China! And drop atomic bombs on the bucktoothed mandarins. But what transpired after that cable is now history. Chairman HO who likes to wear the topcoat of Mother Teresa, and acting as a better strategist than the professional general, ordered instead for the general to Reconstruct Japan, and sent US $Billions of US taxpayers' money to the sneaky emperor whose war crimes as the greatest war-criminal on the planet, continued more efficiently with the additional support by the US 7th Fleet. All the US Aid of $Billions ended up in the emperor's vaults in Tokyo, making Japan as the richest nation on the Earth! The U.S.A. acquired only about eight million body bags filled with Gringo corpses, slaughtered by the Slit Eyes. And the U.S.A. of 2010 A.D. under Chairman HO remained bankrupt with US $Trillions of debts to the bucktoothed mandarins. Of course, the Chairman bought voters, of the Autoworkers Communists Unions, just after Chairman HO's pilgrimage to kiss the Emperor's ass in Tokyo, in a $$2Billion bailout of Ford, GM...for the Chairman's re-election. This cyclic bailout of Ford, GM...will go on indefinitely, because Ford-GM clunkers that are gas-guzzlers and with inferior quality due to an evolutional lag of 150 million baktuns behind the Honda-Toyota made cars that employ automation and robotics to produce the cars, very economically, without vacation leaves, overtime, sick leaves, salary or salary increases, retirement pensions, and without tons of errors produced by unionist morons and monkeys' hands, etc. etc. Of course Ford and GM jacked up cars' prices in the U.S.A. by adding sensors without solving pollution due to unburned fuels in combustion, that Honda-Toyota also imitated, put in under the hood but at least, resolving the pollution issue...all resulting to 375 million car buyers in the U.S.A. (not the 20,000 members of the communist, mob-infested autoworkers unions) paying for the jacked-up car prices as well as the high gas prices. Honda-Toyota cars could have sold in

the U.S.A. at a much less price than the Ford-GM cars due to very little cost in production of the Hondas and Toyotas due to automation and Robotics used by the Japanese...Ford-GM could not retool and, use automation and Robotics because the communists (union members) would lose their jobs! If this is not gross incompetence...I don't know what it is called nowadays.

6. The Slit-Eyed Koreans opened up a war front in Pammunjon to drain more US$ from the US taxpayers. Again Mr. HO handled the war like Mother Teresa. Result: Mexican Stand Off plus thousands of Gringo body bags.
7. Posturing by Taiwan: The bucktoothed Taiwanese announced in the US media, that they will die to the last man, to defend Slit Eyed Taiwan against the bucktoothed mandarins. So Mr. HO again supplied the Slit-Eyed Taiwanese with the latest armaments in the US arsenal, further strengthening Emperor Slit Eyes' arsenal for sneaky use against the U.S.A.
8. The Vietnam War: The Slit Eyes in Vietnam wrested the lands from colonial France, and the Frenchman hollered "Wolf" to Uncle Sam, so bombs were dropped over every inch of Vietnam and Vietnamese (Slit Eyes disguised as Gooks) who burrowed deep underground in tunnels. The bombs were blindly dropped anyway-what the hell-it's only US taxpayers' money! So the Gringos and some gorillas were doped in Bangkok and barely escaped with their spiked asses intact, when the bucktoothed Gooks won the war. Lessons learned in Vietnam: Cry "Wolf" to an uncle who:
   - Wins wars, not just battles
   - Holds on to winnings to share with ALLIES, not with the sneaky Slit Eyes
   - Builds an Empire for the U.S.A. despite all ideologies, instead of building empires for the sneaky emperor of the Sit Eyes or Cleopatra (short for E.T. of the 8th kind).
   - Has far-reaching cognitive development than veneers of Mother Teresa and foresights beyond more than pleasing and getting votes of Man and beasts, every election time!

9. The North Koreans (Slit Eyes disguised as North Koreans) paraded atomic scraps that were given by the Chinaman (Slit Eyes disguised as Chinaman), and then shot several missiles over the sneaky Topknot emperor's

head, who orchestrated the whole shebang in the first place. Result: Mr. Washington who is now completely crazed with nearing election time, proposed, "REARM JAPAN," another US $Trillion proposal to defend the sneaky emperor against the North Koreans (Slit Eyes disguised as Gooks). The US $Trillion was stopped just in time, or is it merely on-hold in the White House?

10. **The bucktoothed Mandarins invaded Tibet. After genocide of the Tibetans and resettling by the Slit Eyes disguised as mandarins, to replace the now extinct Tibetans as present-day Tibetans, the 5th generation bucktoothed Tibetans, rioted against the bucktoothed Mandarins. What for? To get US$ Aid to add to the emperor's vaults in Tokyo, of course. Result One: The Slit Eyed population in Japan decreased exponentially! The only nation on the planet with declining population while the rest of the world has exploding populations, enough mouths to eat the food consumed by all the Tibetan monks since the advent of the biped mammals. Where did all those raw sushis go? All of Asia is now peopled by 5th to 12th generations of bucktoothed Slit Eyes disguised as Chinaman, Koreans, Gooks, Dringo Catholics, Himalayan monks…complete with local names and dialects. And all Asia is now absolutely ruled by Emperor Chop of Tokyo.**
11. **Result Two: Mother Teresa Pelose met with Dalai Lama (the only authentic, live Tibetan fossil on Earth) in Bombay, India to assure the Lama of her support-50 years too late! I can't figure out how Ms. P would support the Dalai Lama; would Ms. P carry the Dalai Lama piggyback across China, over Taiwan…to Hawaii? Or assuming Ms P is now 28 years old, single, and a devout Roman Catholic, would Ms P marry and sleep (in that order) with the live Tibetan fossil to save the only Tibetan (not the fake Tibetans, "Made in China") on the planet from extinction? The Dalai Lama, if he works overtime, could still reproduce a maximum of 22 little dalai lamas by 2020 A.D. (assuming further that the little lamas produced yearly, are always twins of the 1st kind, not the twin of the 2nd kind (Siamese twins) which could mesh up the count, depending on whether the SI Unit system is used (the count of the 2nd kind twin equals two), but in the bloody English unit system, no matter how many pairs of legs and heads a spider twin of the 2nd kind got, the body-count remains as bloody one Spiderwoman).**
12. **The Empire of the Slit Eyes: A year after the sneaky Slit Eyes won the Pacific War, the bucktoothed Dringos**

initially doped Mom at child birthing of my little sister, Lourdes who was killed a few months after birth, and our home initially robbed of cash, jewelries, land titles, titles to cattle…and then Mom was literally bundled to a local gambling den operated by another bucktoothed Slit Eyes…and heavily doped continually for days and months and years, thus leaving no memories of her ordeal (thanks to Dr. Ishii's experiments on live Chinaman at Mukden, the results of the experiments were shared among the Slit Eyes in Asia). For years afterward, Mom was abused. Dad too was doped by the same bucktoothed Dringos in 1949. In both instances of initial doping, only I was with Mom and Dad. 58 years later, my parents perished after being slaves and addicts (without their knowledge)-victims of continual doping, robbing, rape…Without parental guidance and protection, our home and other family members went through the same processes of abuse by the bucktoothed Slit Eyes and their Dringo minions, until 2010 A.D. Two of my younger brothers, a doctor and an engineer (both licensed to practice their professions), are presently languishing in my old hometown, as zombie-invalids after continual doping with heroine, mutilations, and robbing by the Slit Eyes and their Dringo minions, since my brothers' graduation in college in the 1970s. Repeat the same scenario, millions of times all over Asia, besides the rape of millions of Nanking City women, and see what the surviving Asians (not bucktoothed Asians) has become in 2010 A.D.

13. Questions that need answers are: Isn't this Topknot emperor still a war criminal at-large that should have been hung in 1945, along with Togo…? Where did the Slit Eyed soldiers stationed in China, Koreas, and Mongolia…all over Asia in 1944, go? No one was accounted for, after the Gringos' wild celebrations of "We Won! We won!" in the late 1940s. Who are now the small and big landowners in the Philippines, after the Slit Eyed dictator Marcos' Land Reform? The same bucktoothed Dringos remained in possession of their ill-gotten lands after Marcos' eviction from office. And all Dringoe Presidents were Slit Eyes: Quirino, Laurel, Marcos, Estrada…

Greatness Of America

Happy Dreams

Great Patterns

At start of presidential terms, promises made before the election are shelved for the next elections. At end of the term, presidential team members run away with public funds for the incoming official to resolve using taxpayers money to redeem bagsful of deficits from taxes to be collected in the next 500,000 years! Nobody cares how you do your job. Personally, I don't care what methods, procedures and posturing you used. Just do your job right, or I myself, shall impeach you, you son-of-a-beast. Another emerging pattern is first in presidential agenda is to take a pilgrimage to Tokyo to kiss Emperor Chop's ass in order to make clear without any doubts to all the world, that the U.S.A. is a loyal subject nation of the Slit Eyes Empire.

Great Achievements, circa 2110 A.D.

Politicians achieve in getting votes of semi-man (called Negroes) whose brains still focus entirely on eat and copulate like any other beasts, due to an evolutional lag of 150 million years behind Man's brain. The politicians declared by law that the semi-man are created equal to Man due to self-interests and monetary gains, thus transforming the U.S.A. into a zoo to entertain alien tourists from the Milky Way galaxy and neighboring galaxies. The alien tourists are ferried to Earth by the sneaky Slit Eyes disguised as green Martians. 119 million years later, another kind of beast with brains reduced to "pleasing crowds" and "getting votes" emerge through evolutionary adaptation.

The U.S.A. space avionics had closed years before, due to dysfunctional brains after daughters and wives of Man are forced by vote-getting integration laws, also known as Bestiality laws, to copulate with the Negroes. Some Euro nations bypassed the degeneration of Man's brain by herding the Negroes to places apart from Man, deciding to feed them for the next 150 million years till they catch up to present Man. The Tokyo-province U.S.A. becomes a lucrative source of simian hybrids that are more gorillas than Man, for circuses in the galaxies. Another achievement is the appearance of zombies with heads that are always high in the clouds due to rampant suppression of thinking and imaginative creativity. All sorts of preachers, priests, imams...inundate young and

old minds with "holies" through cradle-to-grave drills and practices of fear of holy cross, holy Moses, holy lies, holy ghosts, holy inquisitions, holy mass or rituals, on to the worship of holy cows and crows…Meanwhile, the Russians, the Slit Eyes who are less encumbered with the "holies" have colonized every star systems in the galaxies.

The Great Declaration of the 1960s A.D. "Man=Negroes" had written off, with the flick of a politician's pen, 150 million years of Man's evolution from the biped reptilian dinosaurs; enough time in the evolutionary time frame to reverse Man back to an amphibian frog.

Great U.S.A. Psyche

Military units in uniform that invade another nation are very bad! The sneaky Slit-Eyed military that discarded their military uniforms in China, Koreas, Philippines…and that, in civilian clothes, continued murdering the Asians, robbing the Asians of possessions, lives, names, and identities since 1945 to this day…these atrocities to Man are utterly fine! Before a war begins, a decent Declaration of War must be phrased and then signed by both opponents. Japan's sneak attack of Pearl Harbor was very bad. The sneaky Slit Eyes' continued genocide of 950 million Asians and doping the survivors with heroin since 1945 to 2010 A.D. are very good! In fact, every Topknot collaborator bows to the emperor of the sneaky Slit Eyes, since GW to Topknot HO, as the staunchest ally of the U.S.A. hence erasing into eternal oblivion the eight million gringos and 15 million dringos that were slaughtered by the sneaky Slit Eyes. By the way, how many Slit Eyed soldiers did the gringos kill in the Pacific War? 45,000 give or take 100, mostly consisted of the Slit Eyed sailors who sank to the ocean depths with their ships, and the few Slit Eyes who perished in kamikaze cover ups of the sneaky Japanese hordes. But who were those Slit Eyes wiped out by the gringos in Rabaul, Iwo Jima, Philippines, Hiroshima, Nagasaki…? All were Chinaman, Koreans, Mongolians…disguised as Japanese soldiers, who were forcibly shipped to their doom to the Pacific islands, with nowhere to run unless they could swim like fishes; or as doped-for-decades slaves in Hiroshima and Nagasaki.

The U.S.A. is loved worldwide due to its defense of freedom plus "We Won" all the wars! Anybody from the U.S.A. straying into the Slit Eyes Asian empire, or to the Middle East are kidnapped for ransom or killed outright.

The U.S.A. has NEVER yet won a single war. "We Won" BATTLES but not the wars. Look at Castro, Chavez! We never finish off enemies because when the winning is in sight, the politicians, preachers and priests transformed themselves into jackals after loots, under the habits of Mother Teresa, angels, and saints…hindering us from building Empires. Thus we build empires for others such as Emperor Slit Eyes and Cleopatra (short for E.T. of the 6th kind). So where lays America's Greatness?

It is not in our beliefs in gods and deities because more devout Tibetan monks, who dedicated whole lifetimes to god, became extinct in 1949 under the guns of the Slit Eyes disguised as Chinaman, now disguised as Tibetan monks…Vietnamese monks, Myanmar monks, Dringo Catholics, and Indon Moslems. It is not in U.S.A. leaderships because politicians have led the U.S.A. nowhere but corruptions and repetitive, cyclic bankruptcies. It is not in the great US generals because the U.S.A. would have won all the wars, for all time. It is not in the Negroes (semi-man) that are more gorillas than Man, because Africa would be at the forefront of civilizations on Earth, instead of the "Man-feed-it" state, Africa has always remained since the four-footed gorillas shed body hairs and tails and became bipeds a few million years ago. It is not in commercialized gorilla sports, not in commercialized religions, not in eventually rigged elections, not in all kinds of taxes, association fees, communist labor-union dues (controlled by the mob or Man who become overnight millionaires because of the endless inflow of union-due loot)…

Could it be in Man's advanced thinking brain that gradually adapted and developed to survive in challenging changes of environments during more than 96 million years of evolution? Dwelling in dark caverns, hiding safely from devouring large carnivores, struggling to survive using only hands as bipeds, improvising sticks, bones and stones as tools; finding memory, thoughts, love…atomic weapons to defend family against any aggressors: T. Rexes, pterodactyls, rampaging mastodons, copulate-crazed gorillas, and sneaky Slit Eyed mass-murderers, Topknot minions, and Topknot collaborators! Creativity of Man: the genius of A. Einstein, Nic Tesla, Newton, Galileo, Bruno, the USS Eldridge captain and crew who braved unknown realms in quest of knowledge…deserve at least, street names across the U.S.A.

Great U.S.A. Bungles
Coded messages by the sneaky Slit Eyes were cracked weeks before the attack at Pearl Harbor. The bunglers knew that after Pearl Harbor, Bataan and the Philippines then all the rest of Asia follow as next targets. The main excuse not to act on the information was the Slit Eyes were decently dressed in western suits, therefore, are trustworthy and civilized savages.

After the unconditional surrender by the Slit Eyes signed by the Topknot emperor, the gringos began to celebrate "We Won" with holiday parades and blaring brass bands, and forgot to hunt down and to destroy the Slit Eyes military, INTACT in China, Mongolia, Koreas, Philippines...all disguised as mandarins, monks, and dringos, hence pillages, rapes, mass-murders of the Asian males continued under the military and financial support by the U.S.A. in the US Reconstruct Japan program, the US Aid to Asia and to the Philippines from 1945 to 2010 A.D. And the Asian loots ended up inside the vaults of the Topknot emperor in Tokyo. Why? Because most of the Asian leaders were Slit Eyes disguised as dringos (Marcos, Laurel, Estrada, Sukarno the Indon, Lick One Yu of Singapore...)
Bay of Pigs
Children and infants were burned to crisp in Waco, Texas.
Savings and Loan and other public funds, stolen, resulting to US $Trillion deficit.
Bin Laden's Great Escape to the Japanese Empire in Asia
Cyclic Bailouts of the US auto industry to buy votes for Chairman MaO's reelection, (Autoworkers communist and mob-infested unions) that left Ford, GM...still with an evolutional lag of 150 million baktuns behind Honda, Toyota that had used Robotics and automation, entirely, in car production since the 1970s. Ford-GM could not produce efficient gas-burning, durable cars economically, mainly because using Robotics and automation in producing the cars would lose union jobs and Chairman MaO's supporters. How can Ford-GM's cars compete with the Honda-Toyotas that are produced at very low cost of production because of Robotics that need NO: coffee breaks, vacation leaves, sick leaves, overtime fees, salaries, retirement pensions, errors due to fatigue and workers' handicaps? And Ford-GM's workers' unions jacked up prices of US cars by cluttering the engines with SENSORS without solving the pollution problems due to unburned fuels. Then, the

Hondas-Toyotas also sold cars in the U.S.A. at the jacked up prices. The result is 375 million car buyers in the U.S.A. paid for the 25,000 communist autoworkers union members at Ford-GM to enjoy continual employment. How much did Honda-Toyota get, ongoing to this day, for this scam? 2000% bloody profits, paid by the 375 million car buyers in good old U.S.A. So Topknot MaO is having fun, giving away all these US taxpayers' money to the Topknots of Asia.
The Great "I Shall Return!" by Gen. Mc Arthur
What did the Slit Eyes spend time in during the Asian Holocaust that began in 1939 until 2010 A.D. Chopping the Asian heads off the Asian males, mutilations, rapes and doping of surviving natives, or forcing Chinaman, Koreans and Mongolians to don Jap uniforms to disguise as Slit Eyes soldiers and then shipped to the Pacific islands with nowhere to run. The Asian survivors were doped and shipped to Hiroshima, Nagasaki, and Tokyo...as slave laborers. Gen. McArthur killed them all for the sneaky Topknot Emperor! Hence the Japanese soldiers were disguised as Chinaman, Koreans, Mongolians...in China, Koreas, Mongolia...while the Chinaman, Koreans, Mongolians...were disguised as Slit Eyes soldiers in the Pacific islands of Rabaul, Iwo Jima, Philippines...and the Asian women were disguised as Geishas in Hiroshima, Nagasaki, Tokyo. As of 2010 A.D. we are dealing with 12th generations of Slit Eyes disguised as Chinaman, Koreans, Mongolians, Dringos, Tibetans...complete with native names, identities, dialects. All the misinformation successfully sneaked into the west, with a smile, just like when the Slit Eyes beheaded and murdered their victims-WITH A SMILE!

The Great "We Won"

Wars	Enemy	We Won	We Ran	Enemy Bag	U.S.A. Bag
1700s	English	**		Canada, Australia, New Zealand, Borneo, Falkland	Empty
1800s	Spain	**		Rough Riders; FCastro died fighting the U.S.A. Empire. HMoronChavez also died fighting the nonexistent US Empire. Did the U.S.A possess Spain, Mexico/Cuba, Japan, Asia…? No, No/No, No, No	Empty Bill McK bought the Philippines for $20Million
1900s	Germans	**			Empty-millions dead
1940s	Japan		** Pearl Harbor	Bunglers cracked coded messages of the Japs, weeks before the attack…No action	165000 dead
1945	Japan	**		Bomb China-No! Reconstruct Japan-Yes! The intact Jap military continued mass-murders, rapes, looting and doping of Asian natives.	Empty-millions dead US Aid $Trillions to Asia and Japan flowed into the Emperor's vault
1953	Japan (NKorea)	Mexican Stand Off		Japs against Japs with U.S.A. dancing to the Emperor's claps	Empty
1960s	Japan (Vietnam)		**	Again, the U.S.A. danced to the Emperor's claps	Empty-1000s dead
2000s	Iraq	**		U.S.A. gave away Iraq's oil to Chinaman (Japs)	Empty-1000s dead

	Afghanistan			**Fight for What? Japan.**	
The U.S.A. is barking at the wrong tree; its life style is based on wrong basic principles, or, the U.S.A. leaders are incompetent morons. Why can't we win battles and WARS like other nations on Earth?					

U.S.A. Empire

Reality Check

November 18. 2008 A.D.

1. U.S.A. has been reconstructing Japan since 1945.
2. Pearl Harbor was attacked with 165,000 Gringo casualties, while the Slit Eyes were negotiating terms of war in the White House. In similar manner, the Japanese military, in civilian clothes, that was intact in China, Koreas, Mongolia, Philippines...continued rapes, mass-murders, robbing of the Asian natives, while Emperor Chop's signatures on the Unconditional Surrender documents were still wet.
3. The Japanese soldiers disguised as Chinaman...have subjugated all Asia and built an Asian empire by deceiving the U.S.A. taxpayers who continued sending military and financial aids to Asian nations and Japan, to this day.
4. The woes and imminent collapse of the US economy, are mainly due to these ongoing deceptions by the Japanese Emperor and industrial complex i.e. use of slave labor in China, Philippines...from heroine fed-for-decades generations of Asian natives.
5. Basic complaints by American Negroes are forcible abductions from Africa in the 1800s. The U.S.A. must right these complaints by returning all these Negroes to Kenya, and then clandestinely aid them to take over Somalia, Zimbabwe and Nigeria...all Africa.
6. The Negroes will always lag behind the 200 million-year evolution of Man because of the transitional tree-dwelling gorillas that served as the Negroes' missing link to the tree-dwelling dinosaurs of isolated Africa during the ongoing Break Up of Pangea. (Reference: Any standard Geology textbook with Geologic Time Scale...)
7. A catastrophic major event or two i.e. mega-quake in the San Andreas Fault system, an eruption of the super-volcano in Yellowstone, or a mile-high tsunami could erase the U.S.A. from existence. Only a U.S.A empire could save Man in the U.S.A. from extinction.
8. Changes are necessary, now, that the U.S.A. is still kicking strong!
   U.S.A. Empire
   Goal: To take back all Asia from the Japanese military rule and to create a United States of Earth
   Objectives:

- **Take up by force, if necessary, China, the Philippines, Koreas…all Asia from the hidden military rule of the Japanese.**
- **Return all the loots and booties, the Japanese stole from the Asians and the U.S.A. taxpayers, since 1939 to 2008 A.D.**
- **Return all the Negroes in America to Kenya to stop the Negroes from overwhelming Americans to transform the U.S.A. into the United States of Africa.**

Materials, Equipments: HAAARP, Stealth, Flying Saucers…Everything in the arsenal of the U.S.A.
Standard: Swift, Decisive, Final and Top Secret for 200 years.
Time Frame: November 20 to December 31, 2008 A.D.
Activities:
November 20: Detain and Interrogate Emperor H and sons, all Jap leaders of Mitsubishi, Honda…Space programs… in Quantico to extract by any means, secret codes, passwords, identities, locations…Replace with US personnel.
November 21: Capture and Interrogate all Jap leaders and minions in all Asia, Peru…Replace with US personnel.
November 22: Transport undercover, the first 20 million American Negroes to Kenya to take over Somalia, Sudan, Yemen and Zimbabwe…
November 23: Detain and interrogate all African leadership…replace with American Negroes.
November 24: Tie up all loose ends - no one on Earth must know the changes in leadership in Asia and Africa and the loss of the Japanese Empire to the U.S.A.
November 25: Sample, New Rules
SPalin: U.S.A. Armed Forces, Commander-in-Chief
DickC – US Empire, China, Japan, Taiwan, Singapore, Indonesia
JMcC – US Empire, Vietnam, Myanmar, Thailand, Malaysia
BObm – US Empire, Kenya, Somalia, Sudan, Yemen, etc.

More Patterns

What's Up U.S.A. Man?

Time heals everything, or does it really? Man tends to forgive and forget another Man's past misdeeds as long as there is an obvious change in a current system.

For instance, Saddam Hussein was once an ally of the U.S.A. It wasn't only Saddam who changed but the US policies on armament sales along with the use of such armaments, due mostly to change of a President. Take Kaddafi of Libya as another example. Dictator Kaddafi was once reviled as the Devil Incarnate. For about 60 years, the Devil systematically exterminated all oppositions in Libya, replacing the victims with the Devil's sons, daughters, and the Devil's family members. All lands, resources, inhabitants, oils, minerals...and government of Libya are in the Devil's absolute clutches.

Dictator Kaddafi is among the planet's criminals at-large waiting to be punished, not honored! Likewise the Devil's close friends, Imelda Marcos and husband, who acquired immense wealth during her husband's dictatorship. Lands, cash stashed in Swiss bank accounts, businesses grabbed by Marcos in the Philippines and JapMarcos' Jap followers who remain the owners of those ill-gotten wealth, to this day. Likewise, Emperor Chop of Japan, and any world dictator who like to remain alive in or out of office, after hoarding riches and wealth while holding a position in government of any stripe.

Where is Lincoln's Belief in Equality-in the gutters! Because such members of the society of Man have deprived and robbed millions of Man of the victim's properties, lives, and have subjugated survivors into mindless puppets through doping minds or terrorizing generations of any Man opposing their wills, during their terms in office. Such members of Man's society are the only ones free at the expense of the rest of Man's society. On planet Earth, there is really no place for the likes of these criminals-no place to hide due to a multimedia that uncovers backgrounds and performances of any

Man or any beast holding a position of national leadership.

Patriots, Patriotism! It works; it doesn't work-who cares? Government Service is passed from fathers to sons... After the incumbent's term of office, doing or not doing anything, the officials retire with undeserved retirement loots to France, Italy, Ireland, England...or prostitute themselves as lobbyists in Congress, for Toyota, Honda, Mitsubishi, Shell, BP...and foreign companies. Hence, the present U.S.A. employees in private or government won't have any more social-security benefits, etc. coming to the future retirees, because the premium-deposits for the benefits are wasted as salaries of the officials, as bail-outs for the official's misdeeds in office, etc. etc.

The system works from Prosperity under Top Gun George W of the 1770s till Bankruptcy under Top Knot GW of 2009. The U.S.A. of 2009 is bankrupt!

The three-month Compulsory Service in the government office, if implemented for say, five years, could at least restore the U.S.A. back to the time of Top Gun GW. Then let the beasts rise again to lead this nation into another bankruptcy! The system works for Man, not for beasts.

200 years of trial-and-error in democratic government should have made the U.S.A. the most efficiently run, cleanest, and satisfying government for Man on Earth! Instead, after 200 years, the U.S.A. ends up to total bankruptcy in 2009 and the officials have become very efficient in covering ones' own asses that nobody could pinpoint "whodunit". Why? How? Don't look at me. I've got nothing to do with it. I'm no son-of-a-beast.

What are the U.S.A.'s greatest products-exports? The U.S.A.'s products are: Armaments, Religions with all kinds of preachers and sure-thing government and economic theories toward bankruptcy, corruption...after 200 years or sooner!

ALTERNATIVES

Send more troops to Afghanistan to completely conquer and make Afghanistan, a U.S.A. colony within two years, maximum. If not colonized by 2011 A.D., give up the venture. Leave the Afghans alone! Or drop three ten-megaton bombs (Environmental Friendly), on the Afghans to end all these waste of time, troops, and cash.

Reroute the U.S.A. troops from Afghanistan to Japan, Koreas, China, Myanmar, Philippines, Tibet…to hunt down DEAD Emperor Chop and sons, every Man Jap in Asia and their minions-collaborators. Get rid of all these vermin criminals and build a US empire that 15 Million US GI MAN has given their lives for, in the last wars.

Replace the U.S.A. troops with ten licensed-to-kill assassins-three from the FBI, three from the CIA, and four from the PBA, to hunt down DEAD the unwanted owlhoot leaders in Afghanistan, former criminals like Dr. Kaddafi and his close friend i.e. Imelda Marcos (wife of the Jap Dictator Marcos of the Philippines)…

Why would a die-hard slit-eyed sneaky Jap send his grandchildren to learn and study Mandarin in a Chinaman school in Naga City, Philippines, in the 1970's? Because the bucktoothed, slit-eyed Jap with thick yellowish hide knew that the slit-eyed sneaky Japs were all over Asia, the Philippines, and China, disguised as Mandarins…speaking native Chinaman with Chinaman names and identities…and Tagalog-speaking Pinoys with native Dringo names and identities, since 1939!! And by the 1970's, China and all Asia was completely under absolute rule by the slit-eyed, bucktoothed, bowlegged, sneaky Emperor Chop with thick yellowish hide!

Belly-Up U.S.A.

2009-3009A.D.

1770's Top Gun George W had led young U.S.A. into prosperity never seen before. New immigrants settled on virgin lands and started a new nation with pride and confidence in an ever-bright future. Generations of the settlers worked hard and made great names in the new U.S.A. Then came good and bad ideas, one of which was Equality, Fraternity ...ideas that were influenced mostly, by the Frenchman's invention of the Guillotine that summarily chopped heads of all the great French minds and taller nobility, young and old were not spared!

1800's New U.S.A. became fired up with traditional creation myths and beliefs. Freedom for beasts and Man became the novel battle cry. By blind beliefs in groups of politicians, preachers and priests, the new nation fought each other to free the ex-gorillas as CREATED equal to MAN by numerous dead-end gods! Evolution Theories were then just science fiction, and all reptilian fossils, giant Man remains, etc. that could not be explained by Pope SinBad and its band of Organized Fanatic Thieves, ended up in Smithsonian underground vaults or in the Vatican, never to be seen or heard of, again...

Who put the ex-gorillas in the new land? Vanquished enemies of the U.S.A. Iberians, descendents of older civilizations who had already known corrupt outcomes of mixing ex-beasts with Man; the ex-Beasts overcome Man through the vulnerable daughters and wives of Man, who fall with little prodding, to spread their limbs into bestialities with the ex-beasts... in the name of Women's Lib. etc. etc.

After many shoot-outs wiping out greasers, desperadoes, etc. on land and seas, the U.S.A. gave away Cuba, South America, Mexico, etc. back to the greaser-losers...hence, the U.S.A. veterans' expenses never were recovered with shares of the conquered greaser land. This pattern was repeated over and over again, starting the eventual Belly-Up U.S.A. with nothing to show except millions of filled body bags and US $Trillion red balance sheets.

1900's Wars were fought in many battlefields on lands, in air, and on seas. The U.S.A. military, famously known worldwide as BatMan - because, "Blind as a Bat" they crossed lands and seas, fought, and won BATTLES for

WHAT? (no one knows)! Veterans of previous battles against Man and against hollering dispossessed former inhabitants, called Indians (not from Bombay, India but falsely named by Cris Columbo who after passing Atlantic storms, thought he had finally landed on the shores of India!) of the new world, won all the battles through sheer luck etc. etc.

When slit-eyed Japs clobbered the US GI Man over and over again, since Pearl Harbor, Bataan…then entrenched themselves behind the lands of the now almost extinct Chinaman, Mongolians, and Koreans, disguised as native Chinaman, Mongolians, and native Koreans, while forcibly conscripting the displaced natives to be disguised as slit-eyed sneaky Jap soldiers to every Pacific island, the US BatMan blasted all the Jap uniforms off the Pacific islands, without checking DNAs, killing all the Asian natives in disguise as sneaky Jap soldiers in Rabaul, Guadalcanal, Iwo Jima, Bataan, Philippines, Okinawa, Hiroshima, Nagasaki. But alas!! BatMan could have finished off all the Slit-Eyes on Earth by hunting down DEAD, every sneaky Jap in the Asian mainland of China, Mongolia, Koreas, in 1945 and could have WON the Pacific war!

And Belly-Up U.S.A. continued floating upward! "Enough is enough," Absolutely NOT to finish off the intact Slit-Eyed military in China in Gen. McArthur's "Bomb China," said the homosexual perverts, Top Knot collaborators, politician gooks, preachers and priests. "Reconstruct Japan" with multi-$Trillion US AID and let the sneaky Jap military, INTACT in China, Mongolia, Koreas, and in the Philippines continue raping, robbing, mass-murdering the natives, and doping generations of the Asian survivors into docile slaves, with the BLIND support by the US BatMan, the US taxpayers, and the Top Knot collaborators since 1944 till 2010! And the sneaky slit-eyes and their god Emperor Dracula, amassed ongoing incoming loot and bloody booty from an Asian empire that the U.S.A. has financed and built for the vampire Emperor, for being sneaky and for having slaughtered eight Million GI BatMan in the wars; for continually raping, robbing, murdering 950 Million Chinaman, Mongolians, Koreans since 1939 in a genocide unparalleled in world history; for raping, robbing and doping into slaves surviving millions of Dringo (Filipino) families, since 1945 ongoing as of 2010 A.D. including my BELOVED PARENTS AND SIBLINGS!!!

BatMan jumped into other wars, blindly as a visually blind bat and won more battles but not the wars: Korean War with Japs (SKoreans) fighting Japs in-disguise as NKoreans who are both egged on by Japs (Chinaman) and the Emperor Dracula of Tokyo. BatMan is still there fighting, BLIND AS A BAT!!! Then, came the Vietnam War with BatMan again getting clobbered by the Emperor Dracula's vampires: the Vietnamese gooks (Japs in-disguise as gooks) fighting on stage against the Northern Gooks (Japs in-disguise as Gooks allied to Japs in-disguise as Chinaman) with Emperor Dracula as Supreme Director, just because the Frenchman hollered "Wolf!" to Uncle Sam to preserve its crumbling colonies, including Indo China.

Next is the Iraq Massacre with the U.S.A. spending $Billions and thousands of Man-filled body bags, FOR FREE, for the sneaky Slit-eyes, disguised as Chinaman, to take over the Iraqi Oils.
And the U.S.A. floated up faster toward Belly Up 2009!
The U.S.A. has never won a single war despite the blaring radio commercials and TV- media's large headlines in New York Times, Shinbun, etc. and despite colorful parades in every US city to celebrate "WE WON! WE WON!" the wars!
2000's By the time of Belly-Up U.S.A. of Top Knot GW of 2009, the beasts have thoroughly confused Man's 10% functional Thinking brain, (the 90% are still peacefully asleep and self-hypnotized with daily drill and practice of religious hymns, lullabies, gorilla sports, and Media commercials), so much so that the XX-part simian Chairman MaoHO is now leading Man to climbing trees as tree-dwelling simians! Everyone in the U.S.A. high government gooks are scrambling head over heals over each other, to snatch and gobble down crumbs and bloody morsels dropped from Emperor Dracula's table; forgetting that the Slit-Eyes have unconditionally surrendered to Gen. McArthur in 1945 and had groveled on the ground to spare their sneaky yellowish hides. But of course, the gooks did not finish off Emperor Dracula's military that was intact in China, Mongolia, Koreas, Philippines so the genocide, rapes, pillages, doping of generations of Asian survivors into slaves…by the Slit Eyes continued with more diabolical efficiency, after Dr. Ishii's experiments on LIVE Chinaman, applied to LIVE Asian natives; and the US Aid of $Billions, US RECONSTRUCT JAPAN of $Billions and

Asian loots flow uninterrupted, into Emperor Dracula's overflowing vaulted crypt where the undead emperor exists in Tokyo, ongoing to this day!!

The slave labor used by the sneaky Slit-Eyes in China, Koreas, the Philippines...and the expenses in the adventures of BatMan, are the PRIMARY CAUSES of the accelerated slide-up of BELLY-UP U.S.A. from 1940 till 3009 AD! And the U.S.A. government gooks are still doing NOTHING, not on lowering oil prices, not on emphasizing knowledge instead of commercialized gorilla sports in education, not in enhancing U.S.A. industrial productivity by automation and Robotics...not on fighting and winning battles as well as WARS!!!

Another main cause of Belly-Up U.S.A. is the majority of fanatical cults in the U.S.A. that sell HOT-AIR products, selling like hotcakes to Man and beasts alike. Being hot-air goodies, there are NO AFTER-SALES services, NO WARRANTIES! NO Refunds! NO GOODIES EXCHANGED! The hot-air goodies are for DELIVERY to the buyers' CORPSES, according to the drummers wearing all kinds of sack-clothes, clown-suits with magical chantings, copycat from ancient pagan rituals, etc.

Where do all the $collections-$fees for the Hot-Air Goodies, $donations to the drummers, go, by the way?

The main recipients are (in order of who got the MOST loot):

1. Pope SinBad and its band of 140 thieves, known as Organized Cardinals-Swiss Bank of Rome, Italy.
2. Mrs. King-Bank of England, old Hongkongtown
3. Emperor Dracul of Tokyo-Bank of Japan and All Asia

The only bank that I know didn't drain US$$ out of the U.S.A. is the Bank of Utah!

Finally, CORRUPTION in US government is the last cause. For instance, DOT engineers come out of their holes before every election time, to rebuild roads all over the land. Equipped with the latest heavy machineries and equipments, with a manager monitoring details of the construction process to follow exact specifications, via a small truck with flashing lights on roof and running from end to end of the site for ten hours, all for what? To build the roads to last at least FOUR YEARS or less!

Why can't these highly qualified professional engineers design and construct roads that could have a service life of at least 1000 years or 5000 years, without repairs? The ancient Mayan and Inca savages had done it (the last I've read); had built roads that last 10,000 years all over South America, and used by the Conquistadores to wipe out the Indians (not from Bombay, India…but the Hollywood-type with hook nose, red skin, and riding wild mustangs).

Another example is the Ford-GM Bailouts! What is covered in this bailout is vote-buying (for Chairman HO's reelection) of HO's fellow communists who are members of mob-infested Autoworkers Unions. Do you think that Ford and GM could survive long enough to compete with the high-productivity and high quality of Honda-Toyota cars that sell at the same jacked-up prices (done by Ford-GM due to high cost of employing union workers)? Unless Ford-GM retool and use automation and Robotics like the Honda-Toyotas, and get rid of these communists, in a few more years, another Ford-GM bailout would again be given by the Chairman, disguised as U.S.A. Economy Bailout, to again, buy the votes of the Chairman's fellow communists. Honda-Toyota profited 2000% from these communist unions' control of Ford-GM. Why? Because Honda-Toyota produce cars, employing Automation and Robotics that need no coffee breaks, salaries, overtime fees, retirement pensions, vacations, sick leaves, etc. sticky hands of communist morons of the unions? McC and P Patriots! Impeach Topknot MaO now, for GROSS Incompetence etc.

Then there is the well-known fact that to get hired in a good-paying job depends on WHO ONE KNOWS, another way of saying how corrupt the system is i.e. in the public-school system for teacher applicants, and any other good-paying government posts; jobs of parents passed on to children are true anywhere in the U.S.A. systems. Is the U.S.A. a monarchy or well-organized mob-infested Labor UNIONS of despotic Mao communists (Japs in-disguise as Mao Commies)?

The SOLUTION to solve this thousand-year Belly-Up U.S.A. is very simple. Capture NOW, the sneaky Emperor Dracul of Tokyo and all the manJaps in Japan, China, Koreas, Mongolia, Tibet, the Philippines… and give them two choices: die now by hanging or by Flaying their thick yellowish hides! BatMan U.S.A. has the right to protect and defend U.S.A. Man from these accomplished sneaky Slit-Eyed recidivist-criminals, Jap collaborators and Jap minions!!

Try also making these gorillas that are mostly on government welfare, produce something useful, by transplanting Man's thinking brains to replace simian brains that focus entirely on copulate and eat, reproducing more gorillas to overwhelm the U.S.A. into the United States of Africa.

Then control the sale of the HOT-AIR GOODIES or make all the hot-air goodies as illegal goodies like marijuana, heroin, cocaine, etc.
Unless these primary causes are neutralized permanently NOW, the U.S.A. will remain Belly Up for a thousand years, and will be enslaved by the Asian empire of the sneaky Slit-Eyes of Japan led by Emperor Dracul.

Where are the promised Changes, etc?
Do you like the U.S.A. to wait for 75 Million years more, before Chairman Mao Half-N-Half makes the changes?
THIS IS A CRY FROM THE LAST U.S.A. PATRIOT, FOR UNIVERSAL JUSTICE!!!

It Aint Right

By Nic N'Goya

I'm impeaching Chairman Mao and Dick Nixon.
Why?
First, copycat HO is stepping on Nixon's tracks by hastily scampering to Emperor Dracula's summon to promptly appear in the Undead's Imperial Lair in Tokyo, like any other Top Knot minion, thus heaping indignity to the august Oval Office. Instead of just sending Enola Gay to drop its load on top of Top Knot Dracula's head, Mr. Ho groveled with bowed forehead touching Dracula's foot, in symbolic unconditional surrender of the whole United States of America, as a subject nation, to the slit-eyed Emperor Dracula's Empire, without a single shot.

Then, Mr. Ho shared Emperor Dracula's table and supped, without throwing up, slimy raw sushis, that were sliced thinly from boneless rumps of 950 million headless Chinaman, Koreans, Dringos...while the 950 million severed Asian heads bounced like basketballs, all over the Imperial Lair, screaming voicelessly through wide-opened rictus of a mouth, in helpless protests at being untimely severed from their bodies, by the slit-eyed Emperor's Kempetais.

How can these dual monstrosities feast with Ho's happy-chimpanzee smile and both grinning from ear to ear, with crimson blood dripping down their chins, amidst nightmarish surroundings such as these? After the grisly feast of raw Asian-meat sushis, both Mr. Ho and Emperor Dracula swam squealing with fiendish delight, splashing each other with bloody seawater on the Sea of Japan that was filled with the 950 million floating headless cadavers of the rotting Asian bodies.

And both monsters parted with tears in slit-eyes and round chimpanzee eyes, not wanting to end playfully draping and whirling the headless corpses over their heads. Before parting, the Undead Emperor declared as a parting memorial, "Goodbye Ho, my only adopted son-of-a-beast, with whom I am well-pleased."

Mr. Ho continued in a fast clip to meet in state-visit his close comrade-Hu Jintao, the tenth-generation Jap disguised as Chinaman, in Beijing.

Like true samurais, cold-blooded Ho and Hu walked hand in hand, passing Nanking City on their way to Beijing, seemingly unaware of the millions of nude dying and dead daughters and wives of man: mothers clutching little girls protectively, with their bloodied spread-eagled limbs and partially eaten inner thighs that were crawling with sneaky slit-eyed bowlegged sperms, wriggling out their bloody...
"Que los diablos pasa aqui?" Liar Ho asked innocently Hu, without looking at Hu's slit-eyes.
"Nada, no hablo Englis." Bucktoothed Hu replied in mandarin, slit eyes avoiding Ho's round chimpanzee eyes.

Every city, from Nanking, Peking, Chungking, Hongking...were strewn all over with ravaged and spread-eagled naked corpses of hapless female Chinaman, Koreans, Dringos...mothers, girls, old women, every one repeatedly raped by the Jap rapists and genocidal maniacs, since 1939. Then both Top Knot minions happily hopped to Mukden Camps that are still operating all over China, Koreas... full-blast, non-stop since 1939!

Mr. Ho with Jap Hu then, inspected and saluted the endless lines of slit-eyed Jap military that was INTACT in China since 1945, all disguised as Chinaman and ready to sneak into the good, old U.S.A. under Mr. Ho's guidance. IT AINT RIGHT!

Mr. Ho danced back to the Oval Office, triumphantly displaying his accomplishments that no Man or Kabuki had done before. Still reeking with the stench of the undead Emperor and the rotting corpses of the 950 million Chinaman, Koreans, Dringos, and eight Million Gringos slaughtered by the Slit Eyes since 1940, Mr. HO with hide glowing proudly with his Dynamite-Stick Peace Prize, presented to the U.S.A. Man, a successful negotiation of another Pearl Harbor with the Slit-Eyed Sneaky vampires!

Barking Up A Wrong Tree
Great Wall of China at Rio Grande

Man would continue barking up a wrong tree unless one puts a stop to the Man. It's like the popular story of the Emperor's Clothes, "Oooohhhh! Aaahhhh!" hollered by ALL, except a young spectator who ejaculated, "HE'S NIKKED!"

To earn a living, Man would act most outrageously to blindly follow employers' orders and policies; sacrifice personal principles, be hypocrites…just to keep one's job. This is true to Man who professes lifetime devotions to dead-end deities, be it Christianity gods, Moslem's gods, etc. etc. Dump and shoot your neighbors or is it "Love Canadian or Mexican bipeds as Thyself" It's a nightmare!

Texans (I'm a month-old Texan, migrant from Florida), once upon a time enjoyed freedom to ride anywhere in Texas or Mexico, as long as one has a horse. Saddle up and hit the trail! Anywhere you like: north to El Paso, on to Kodiak Alaska…or south to Corpus Christi…and one can stop on every saloon along the way to imbibe poison: Tequila, rotgut or what was there to quench a thirst. Afterward, one sobered up on the saddle, while the horse took him home.

There were no checkpoints. No one would ask questions: why you look silly, where you came from, or where the hell, you're going? Those were the happier days for the Texans.
Nowadays, when a Texan likes to ride a wagon to the nearest Wal-Mart store in El Paso, to buy monthly supplies…there will be "Nicaraguan" checkpoints along I-10 East, nightmarish interrogations, etc. especially if the Texan happens to come from a third-world country where those in uniform are paid mercenaries by dictators, sneaky Jap drug lords…and are the only armed-followers of the "law of the gun".

It is a good idea to preserve the neighbors' goodwill and prevent illegal entrants along the borders, instead of developing animosity, BUT HOW? The answer is the good idea, more so if revenues into the state's coffers continue inflowing to pay for the "job-creator" border-patrol agency.

This is where the Great Wall of China comes in. Would it please the Canadian or Mexican bipeds to be walled in or walled out (depending on where you look)? I personally don't think so, too.
How about making the US border-patrol the greatest show on Texas, so foreigners and Texans could enjoy watching the border-patrols at work? Instead of chasing away the foreigners-tourists with nightmarish checkpoints along I-10 East, why not give them a show along the banks of the Rio Grande, so the tourists would, most likely return to Texas, over and over again.

This is an easy task for the agency.
Firstly, change the patrols uniforms with gold and blue with yellow or red stripes on some strategic places to catch the beholders eye: gold epaulets, yellow armbands, knee-high brown shiny boots with clanking six-inch silver spurs, a blank-bullet studded wide leather belt buckled on the patrols hips, with a pair of bone-handled peacemakers in tie-down holsters, a lever-action Winchester rifle, a wide-brimmed Stetson Texan sombrero and a saddled horse grazing nearby (to chase and run down any four-footed Mexican illegal entrant all the way to Idaho) would complete the patrols' accouterments.

Secondly, divide the whole length of the borders along the bank of the Rio Grande into three-mile sections, with each section patrolled by at least four full-time border patrols, marching in precise military cadence on the section so that every inch of the whole three-mile length along the Rio Grande, is always in view by at least one border patrol.

Thirdly, remove the checkpoint on I-10 East and make it a deluxe rest area for tourists and drivers. Then build rest areas overlooking the Rio Grande every ten miles, for the tourists to take videos, pictures...of the greatest marching show along the banks of the Rio Grande.
With the blank bullets in the peacemakers and rifles, no one would accidentally shoot one's foot, Canadian-Mexican bipeds, or four-footed Mexicans illegally entering the U.S.A.

Fourthly, divide the day into eight-hour shifts so the great show on the Rio Grande could be a continual enjoyment in sunshine, sunshine, and sunshine with a little shower once every three months, 24/7!

WHY NOT?
If the border patrols' productivity is consistently insignificant PERMONTH, or, ZERO number of drug-cartel criminals, ZERO number of Osama terrorists, ZERO number of illegal entrants, etc. in spite the 100 % inspections and questionings of each driver passing the checkpoint along I-10 East, these indicate that all these present operations by the border patrols, particularly at the I-10 East checkpoint, aren't effectively achieving the agencies purported objectives. Besides, sort of harassing Texan taxpayers, foreigners and tourists so as NOT LIKELY TO RETURN to Texas, hence reducing substantially incoming revenues to the state treasury, the suggested modifications could transform the Texas Border Patrol into a peace kind of thing and very neighborly, creating goodwill instead of building up enmity.

Lastly, build new public libraries in every Texan town, including ghost towns, and hire more librarians! Why should Texans be brought back to the Dark Ages, by "reburning the library in Alexandria"?

Carbohydrates

NicWBriones Honors10 Biology

12-06-07

Objectives:

1. Define Carbohydrates, types of Carbohydrates, and give examples of each
2. Know approaches to problem-solving and reading for comprehension
3. Develop long-term memory

Materials and Equipments:

1. Handouts: Preview Chart, Carbohydrates article, Venn Diagram
2. Laptop with Projector and Screen

Instructional Strategies:

1. Teacher-Mediated Learning, Interactive Discovery
2. Problem-Solving, Cooperative work
3. Discussion

Activities:

9:00 Pledge/Announcements
9:15 Bell ringer/Attendance Roll
9:30 Introduce Carbohydrates/Check Prior Knowledge
9:45 Distribute Handouts/Give Directions:

- Complete first, the Preview Chart
- Read the article for comprehension/you may take notes
- Model some approaches to solving the problem i.e. Brainstorm

10:00 Work in trios to solve collaboratively, the given problem.
Be systematic/sequential.
10:30 present your solution to the class, for further discussion.
10:45 summarize the lessons (Venn Diagram)
11:00 End of Period

Assessments: Informal Observation, Formative/Summative Questioning

SSS: SC.F.1.4: SC.H.3.4.5

Problem: Mr. Doe and Mr. Jake agreed to try high-carbohydrate diet for 3 months. They will meet after the 3 months to assess what happens.

Three months after:

Date	10-12-10	01-12-11
Mr Jake	120 lbs	300 lbs
Mr Doe	120 lbs	120 lbs

1. Brainstorm for list of relevant ideas from the

articles. (Whole class)
2. Discuss the ideas in-group of 3 students, and construct a daily agenda of 10 items.
3. Present your agenda to the other groups.
4. Discuss and make an appropriate daily agenda

Sample Solution:

Brainstorming List of Ideas	Mr. Doe		Mr. Jake	
1. Initial Weight: 120 pounds (Given)	1. 120 #		1. 120 #	
2. Carbo-Loading (Given)	11		5	
3. Intensive Exercise / Physical Activity	5		(3)	
4. If taken before=Hypoglycemia	3		(8)	
5. Break back glycogen to glucose; use as fuel	8		2	
6. High-Glycemic Index food	2		6	
7. Store excess glucose as polymerized glycogen	6		(4)	
8. Within 15 minutes - 2 hours after	4		7	
9. In liver and skeletal muscles	7		9	
10. 120/300 pounds (01-12-11 Given)	9		11	
11. Sleep 8 hrs. No drugs, nicotine	10. 120 #		10. 300 #	

COMPARE-CONTRAST CHART
CARBOHYDRATES

DIRECTIONS: Use the information from the article to complete the chart below. First, tell their common characteristics. Then write down their differences.

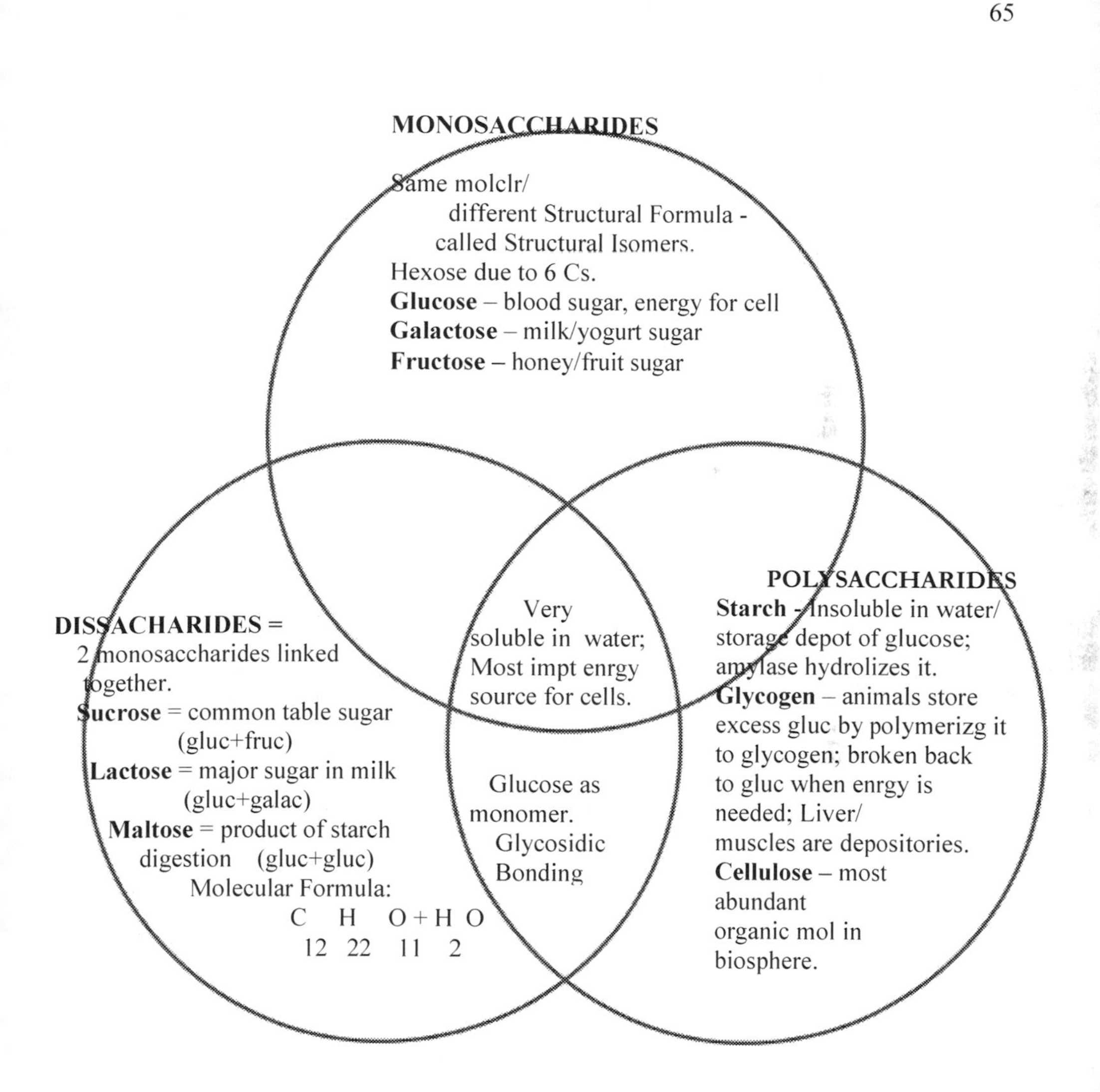

PREVIEW CHART
CARBOHYDRATES

DIRECTIONS: Complete the following chart before reading the article.

Title	Change the title to a question
Read the introduction	List the main points

Read the headings and list them here	Change each heading into a question
Read and list the words that are underlined or in quotation marks	Why do you think these words are emphasized in the article?

Survival City, 2109 A.D.

City of Man, Pop: 50 Million

PURPOSE: Survival of Man under continually worsening environments and under reverse evolution, due to bestiality and undue respect for obsolete theories, beliefs, religions, etc.

To properly direct forward evolution of Man instead of reversal to beast hood, by not giving Bestiality Rights to ex-gorillas and ex-apes to copulate with daughters and wives of Man.

FREE: Shelter, Food, Clothing, Education, Healthcare and Power...

What motivate Wars, Crimes, Drug Cartels, Corruptions, Dictatorships, Prostitutions, etc.? Hoarding of Money, Possessions, Properties, Fanatical beliefs in deities, preachers, priests, cults and religions...

POLLUTION: Mainly from Unburned Gases from Internal Combustion Engines, Coal- or Oil-Fired Power Plants as well as from radioactive wastes of Nuclear Power Plants.

ELIMINATE:
Transportation in cars, trucks, ships...Production, Sale, and Use of Internal Combustion Engines.
Toxic pollutants from Internal Combustion Engines, Coal- or Oil-fired, and Nuclear Power Plants, waste Paper, Plastics, glass and Non-degradable materials.

1. Unclean and toxic Rivers, Lakes, Oceans due to wastes from Mines, Industrial Processors etc.
2. Extinctions and contaminations of Aquatic and Land animals, fish, and wildlife etc.
3. Deforestations, Liquefaction, Landslides etc. due to concreted pavements, blacktop over roads, highways, parking lots, commercial malls, residential subdivisions, government and religious building complexes, schools and universities etc.
4. Floods and erratic inclement weather patterns
   Food Preservatives, food processing, pasteurization, irradiation etc. to prolong shelf life, and hoarding.

5. Destruction of environment, ecology, wildlife…and the Earth
6. Non-essential and no longer functional government personnel, politicians, etc.
7. Waste in use of resources, power, the Sciences and Technology due to restrictions of traditions, nonfunctional laws, commercialization, etc.
8. TAXES and Non-functional government agencies and bodies
9. MONEY
10. CRIMES due to malnutrition-retardation, poverty, insanity, greed, ignorance, gangs, cults, religions…

SHELTER
Diameter: Minimum three miles
Height: Minimum 900 feet
Fireproof, Ventilated for comfortable Temperature and Humidity
Water Storage in underground tanks
Mobility: Bicycle, Walking, Jogging on outside perimeter
Multimedia available in libraries for news, schooling, entertainment…
Maintenance: Volunteerism of three months per year

COMPULSORY SERVICE: Three months per year inside or outside the City by sign-up roster of voluntary candidates based on first-sign first-serve basis.

FOOD
One Kitchen, for economy, safety etc.
Fresh food products with no long-term storage and no refrigeration;
No preservatives, no pasteurization, no processing, no packing…
RECYCLERS
All waste products: papers, plastics, organic and inorganic materials…
REVENUES: Three Months per year compulsory service, not paid in gold or Money; Volunteerism of not more than three months per year.
POWER
Solar Panels and Battery Storage on top dome
Wind Power and Battery Storage on walls, facing wind source
Geothermal

TRANSPORTATION, Mobility
Dirigibles: Solar Panel, Battery powered; Bicycles

SAFETY
No pets: dogs, cats, beasts of all kinds, homosexuals, ex-gorillas
No drugs, deadly weapons, martial arts, gangs, cults, religions, priests, preachers
SECURITY-SAFETY
Assignment in required three-month compulsory service plus Volunteerism

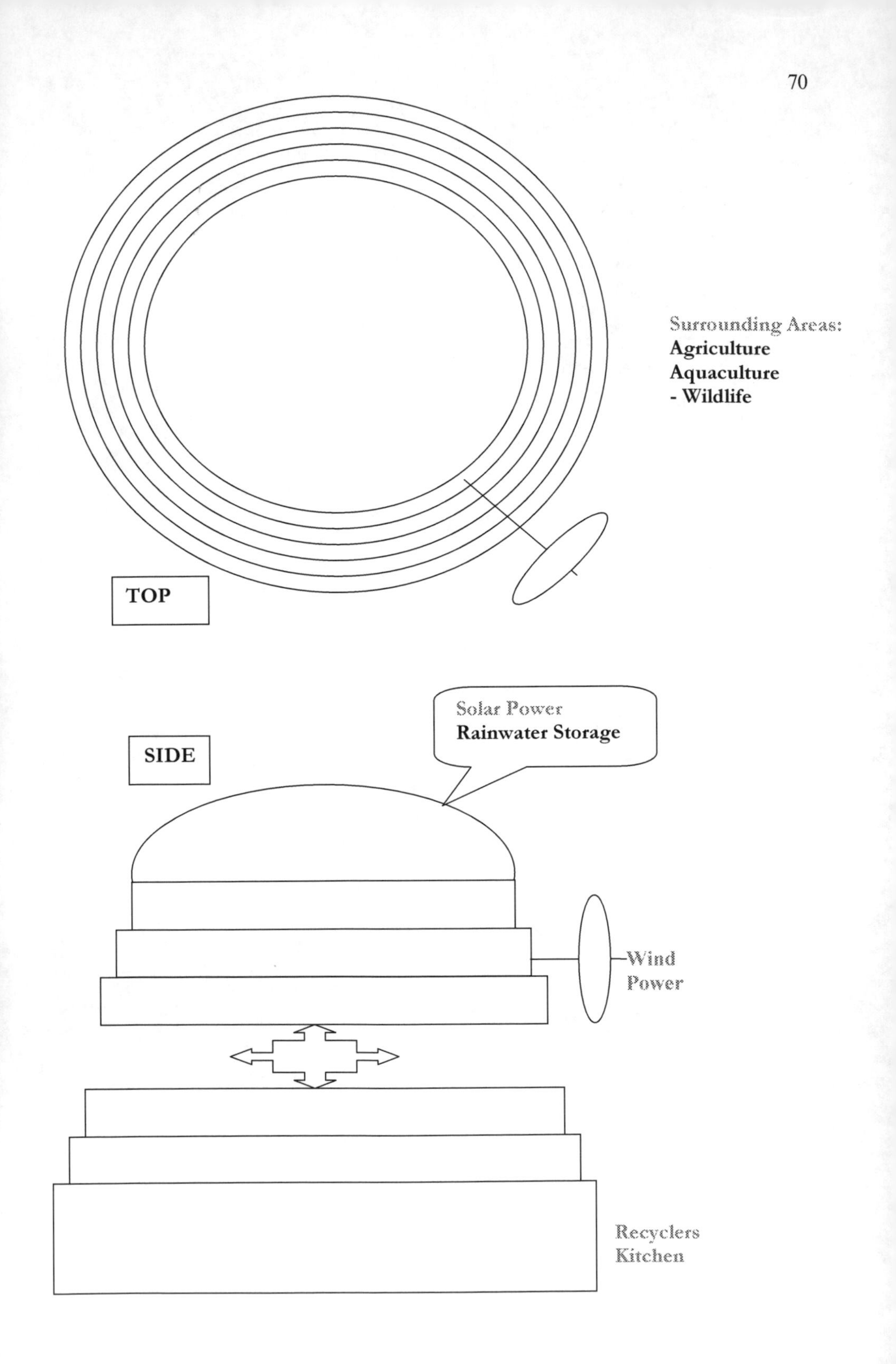
Surrounding Areas:
Agriculture
Aquaculture
- Wildlife
TOP
SIDE
Solar Power
Rainwater Storage
Wind Power
Recyclers
Kitchen

WAIT TIME ONE

Buzzards

Ever observed vultures perching and waiting motionless on tree branches? To find out why the buzzards patiently sit on the tree branches ... throw a dead cow on pastureland near the tree. After a moment, look up and watch a high-flying vulture scout, circling around the pastureland. Then another vulture joins the first, after another moment there are three, four ... many buzzards circling in the skies above the dead cow. As more vultures join the buzzards above, some buzzards begin to land on the pastureland.

A dead-cow feast begins. Non-stop feeding sounds of frenzied slurping and tearing flesh go on till the last bit of carrion is stripped from the cow's bones. And the vultures could only waddle aground, too filled with meat, too heavy even to walk far – grounded for some time.

Days and weeks after the feast,
The vultures on the nearby tree branches perch, like black flowers, watching with shifty red eyes;
For another dead-cow feast to come.
No more dead cow; away fly the vultures after a few weeks of wait time.
If a dead cow is thrown everyday, for weeks, for months, for years, for 200 years on the same pastureland,
How would the vultures behave?

WAIT TIME TWO

Jaws and D'kangs Forevermore

By Edgar Allan Poop

In D'kangland far away, deer, sheep, fruits, nuts abound; bubbling spring cool drinks provide to 60 million healthy, robust d'kangs. Countless palm-frond shelters along endless seashores, dot tens of thousand villages, d'kang towns.

The happy d'kangs, every weekend have fest, singing and round dancing; a male d'kang holds hands Annabellee d'kang, on hind legs straight both closely stand, preen round and round ... The singing and dancing end when, the male d'kangs raise up hind leg and love spray d'kang Annabellee. Dance now, marriages later produce little baby d'kangs in the palm-frond shelters, countless prosperous towns, 60 million d'kangs strong, In paradise of D'kangland.

One silent night, no moon no twinkling stars, the d'kangs' keen noses catch whiff from far: Danger Coming! Doom scent in air! Inside the palm-frond shelters of the d'kang villages and d'kang towns, instant howling at no moon starts. Li'l d'kangs' wailing fills the silent night no twinkling stars.

120 days away, in fleet of sails, a Cook captain asks Nile crocodile DONDE? The crocodile's reply,"Bloody rare, on-the-hoof, still kicking," grill-choice by a million Israelite-Egyptian veterans. The million veterans, King Moses led, sail in the fleet 120 days far. Four months, swift sail unfurled, cross-vast oceans ... at last the Cook captain sees land. LAND HO! Land Ho! King Moses and One million Israelite-Egyptian veterans - not far behind, land at endless seashore of D'kangland!

King Moses prays, Final solution how?

Holy Cow! King Moses caws like a crow, "How how, now now, cow cow?" Cow god thunders reply.

KILL! KILL ALL!! KILL ALL!

Wipe out all GIANT MAN, prosperous neighbors Egypt 'round. Kill all fathers, mothers with child in arms, boys, girls, and old ones.

MERCY, SHOW NONE! Only King Moses' fish ears, thundering voices heard, cow god's commands.

Be NICE GRABBERS of GIANT neighbors' lands;

RUSTLE all livestock: cattle, sheep, goats, and winged fowls;

All grains, all milk and honey, all GIANT MANS' possessions and properties:

STEAL! ROB!

HOW ELSE one million Israelite-Egyptian bandits BUY horses, chariots, swords, shields, and lance?
HOW ELSE one million Israelite-Egyptian LIARS, Feed and Survive FOR 40 YEARS ON DESERT barren LAND?
King Moses, Mastermind, leader of the Israelite-Egyptian bandits, did as his filthy ears alone, cow god's commands heard;
And writ in book of HOLY LIES With silly grin from ear to ear, Big Mac, Pepperoni pizza, French fries, from heaven Mars? Venus? Fall out of thin air! Manna from heaven falls for 40 years, to fill wide-open stinking mouths with rotten teeth of cow god's chosen: one-million Israelite-Egyptian bandits, rustlers, manic mass-murderers, congenital liars!
In D'kangland, King Moses ordered one million Israelite-Egyptian bandits, as cow god's commands
King Moses' ears alone, god's voices heard.
KILL! KILL ALL D'KANGS!
KILL ALL D'KANGS!
Kill fathers, mothers with babe in arms, young ones and old ones. MERCY, SHOW NONE!
STEAL! ROB! RUSTLE! RAVISH! BURN!
All D'kang villages, all D'kang towns,
RAZE TO THE GROUND!
NO GRAVES, NO HEADSTONE MARKS, NO SUN-BLEACHED BONES, NO TRACE OF CRIMES!
As cow god commands, King Moses' ears alone heard; the king of liars, thieves, mass-murderers...obeyed and did,
THE PERFECT CRIME!
By day, the Cook all crocodile smiles, barters glass beads for gold nuggets, diamonds, rubies in d'kang villages and d'kang towns.
By silent night, no moon no twinkling stars, King Moses and owl hoot band,
Swoop down on d'kang villages and d'kang towns, like the buzzards on dead cow!!
SURVIVING d'kang captives bound, line single file, double-step march to Music sound "Auld Lang Syne".
To "walk-the-gangplank" grilled bloody rare, on-the-hoof still kicking:

Fathers, mothers with babe in arms, young ones and old ones; one-by-one drops at gangplank's end; Fall into JAW'S wide-open maws.
ChompChomp! Razor-sharp teeth chompchomp.
Hmmmm, Yummy yumm, bloody rare grilled D'kang meat, down JAWS' gullet one-by-one.
No graves, no headstone marks, no sun-bleached bones, no give-away trace of perfect crime.
WAILING D'KANGS, GOODBYE! FOREVERMORE GOODBYE! SEE YOU LATER, NO MORE!
LI'L ANNABELLEE D'KANGS!
Hour after hour, day after day, week after week, month after month, year after year, for 200 YEARS!
Li'l D'kang captives bound, line single file, double-step to "Auld Lang Syne", walk-the-gangplank one-by-one;
Drops at gangplank end, Fall into JAW'S wide-open maws, Chompchomp razor-sharp teeth chompchomp.
Hmm yummy yum slides down JAWS gullet.
No graves, no headstone mark, no sun-bleached bones, no trace of perfect crime!
NO MORE D'KANGS! NO MORE LI'L D'KANGS! FOREVERMORE!
Swimming JAWS still, at exact spot below, walk-the-gangplank above; hopeful for bloody-rare grilled D'kang meat.
WAIT TIME, FOREVERMORE!
The last D'kang captive bound, gone chompchomp, hmm yummy down JAWS gullet…
Goodbye d'kang Annabellee!
GOODBYE FOREVERMORE!
Moses, King of Liars, with silly grin from ear to ear announcement makes:
I only did what cow god commands!
Mine ears alone, thundering god's voices heard!
Loose screws and rattling nuts inside King Moses' head:
A burning bush, no fire, no flame, "Only mine eyes alone hath seen."
Turn wooden sticks into vipers, with forked tongues!

Divide apart the sea.
Manna falls out of thin air from heavenly Mars? Venus? To fill gaping stinking mouths with rotten teeth of a million Israelite-Egyptian congenital liars, thieves, rustlers, manic killers, for 40 years! On desert barren land!
After filthy deeds, after filthier deeds and unspeakable crimes;
Rename Israelites-Egyptians, Hebrews;
Re-rename Hebrews, JEWS! Re-re-rename JEWS, Messiah man! Moslems! Sunnis, Baptists, papal Cathols!
STILL REMAIN THE SAME,
Israelite-Egyptian thieves, congenital liars, rapists, genocidal maniacs!
FOREVERMORE!

"OWWL, HoOoOOot!!
OWLLLLLLL, HOooOT!"
Cleopatra (short for E.T.) hooted.
In English brief: "Cook captain, all owl hoot bands! Flip up Filthy ears, Hear! Friends, Romans!
Israelites-Egyptians thieves, murderers, liars! Countrymen!
From D'Kangland To old Hongkong town, Bigfoot no more!
Or chopchop Heads! Owl hoot Heads Off, FOREVERMORE!"
"Owwwwwwllll, HoOOot! Owwwwllll, Hoot!"
The owl hoot queen, hooted no more, FOREVERMORE!!!

THINKING BRAIN OFF, the brain shrinks 119 million years later.
Lies and faith on lies, few surviving Man wallow as beasts, in faith on wicked Lies!
Lies on Liars lips remain LIES not truths, repetitions ten trillion times, from cradle to grave –
LIES REMAIN LIES NOT TRUTHS, FOREVERMORE!

King Moses THREE FANATIC cults found.
The JEW cult, the ALLAHU-MAN cult, and the MESIAHMAN cult.
Each cult believes as the ONE AND ONLY CHOSEN,

By each of THREE NOT LOOK-ALIKE CREATOR GODS!
The JEWS see the Allahu-man's head as not right, with loose screws and rattling nuts inside!
The Allahu-man sees the Messiah man's head as not right! With more loose screws and more rattling nuts inside!
The same insights, the Messiah man sees both JEWS and Allahu-man...

SO each cult hands around each other's throats, squeeze shut windpipe, out air not in.
ALL HOLLERING:
BREATHE NO MORE! CHOKE! DIE!
Die Allahu-man Die! Die JEWS Die! Die Messiah man Die!

All cults believe own cult as
ABSOLUTE! EXACT SCIENCE!
No questions ask, no thoughts and doubts allow, no change of words on Cow's book of holy LIES!
THINKING BRAIN COMPLETELY OFF!
SHRINK!!
OBLITERATE till MAN'S Thinking Brain,
A LOOK-ALIKE, BEAST'S ZERO BRAIN!!
All the cults need, ten percent of income,
NO LESS or the BETTER MORE,
As membership tithe,
To receive god's heavenly rewards –
AFTER ONE DIES, TO A DEAD CORPSE!!
Three heavens await the dead corpse: First heaven full of beautiful belly dancers - ravish all the dead can devour!
Second heaven filled with limousines, $$$ and Gold, wines, mansions, and mistresses on the sly!
Or filled with bloody burning HELL, for dead Man and dead beasts not able to pay the ten percent of income tithe;
Where only devils with blistered skins, two bloody horns one long red tail with arrow at an end,
Thrive in Lucifer bliss, in fiery heat, all year around.

WHO LIED THE BEST?

Basket-baby Moses! Manger-baby Messiah
man! Mark with seven-pointed star high in the
skies
Instead of a round sun, to guide magical three gift-
bearing mid-eastern kings…
Virgin-Born by Mother Teresa!
King of Liars parted apart the sea and walked on dry
seafloor. Mesiah man simply walked on the waters -
What the hell!
Moses' manna out of thin air, falling from where in
mapped heavens, WHERE?
Mesiah man fed 5000 liar snouts,
From Magic basket spewing magic fried fish and
bread Loaves;
Plus Mesiah man : Water instantly brewed to wine, in
wedding magic jar.
Plus Mesiah man, as the BIGGEST Liar of all:
REVIVED a play possum Liar, back to Stinking liar
LAZARUS!
The Messiah man thief was killed by Romans. After
three days as smelly corpse, reanimate, and walked
Earth for magic 40 days,
Then like rocket ship to heavens where, even NASA
can't specify –
As writ by a LIAR Jew whose magic two eyes alone,
nobody else saw all.

For salary plus housing free, for livelihood, for meal
ticket, for laziness,
Slurp, chomp and slurp Campbellees work with spittles
drool,
Out Liar's snouts
Holy Cow's words, holy cow's Lies;
From king-of-liars holy book, from Moses holy snout,
Preach Vilest Lies, not truths.
Dumb A-Holes, papal Cat-Holes, Allahu's Oil-Holes,
LIARS' SNOUT HOLES!
Preach lies, lies, more lies; Preach vilest Lies
FOREVERMORE!!
Accepted Hard Evidences of Lies as truths:
Believers with university Ph.D.s, Fanatic members
of the cult!
Other cult fanatics are Presidents, senators, governors,
Kings, Queens, the richest, and mostly the ignorant,
poor Man on Earth!
Dutchman, Frenchman, COPYCAT MAN, some
Orientals…Cult members all!
5000 years, cradle-to-grave repetitive drills and

practice of holy lies, AND unspeakable crimes:
from burn at stake "witches and heretics",
Millions of natives wiped out, in foreign lands…
LIES REMAIN LIES NOT TRUTHS!
FOREVERMORE!

studying in Feati U. Salvador was and is still a very active leading member of these dopepushers-robbers who also doped Dr. Jun every night, in my sight since 1973 during Jun's doctoring practice in our old hometown, Ocampo, Camarines Sur, Philippines until 1981, when I left for Manila to get my immigrant visa to the U.S.A.

When I left Ocampo in 1981, the Bombase brothers stood outside Oscar Asetre's house on San Isidro Street, and looked at me sneering at me with laughter, shouting in a very loud clear voice: "JUAN BOMBASE JR! DOMINGO BOMBASE!" several times. This was as good as admitting their direct involvements in the doping and robbing of our parents since 1945, and Dr. Jun and Engineer Achil…in Ocampo since 1973. And a few days before my departure, Ricardo Martines in uniform as policeman, entered and sat in our house. When I entered, Ricardo put his hand on the butt of his pistol, sneering at me and daring me to do "anything".

I may have neglected caring for my health due to youthful ignorance and lack of adult care (both our parents were already disabled by long-term dope addiction, since 1945 also without knowing they were addicts – not even by us, due to the gangs' cover-up misinformation spread specially to our family members, was effective to our young minds, that Dad was merely "drunk" with alcoholic drinks), but I've self-taught myself about life and developed my thinking brain through reading and movies, and I was the only member of our family who had completely severed physical proximity to these loathsome dopepushers, robbers, and liars since I came to Manila. And I was also the only one who had returned and lived with our parents for 12 years, as an adult after finishing schooling, with knowledge and a feeling that something evil about this gang, was connected to events in our home and family.

Hence I've observed and listened to our parents during their "high-with-dope" periods with Dad loudly talking as if "drunk-with-gin…" though not smelling of the alcoholic drink. I've also watched helplessly, this gang

CRIME PAYS

Just So You Know

(A letter to some close relations). How're you all?
I'm feeling great – no more rushes to meet schedules at work, schools…I still watch movies, read or listen to audio novels, when I can. I'm also hoping to teach and earn better income…so I could help you to all come here.
What follows will appear incredible to you; you might even refuse to believe me.
The Japs and their Filipino minions have overrun the Philippines and the Japs have subjugated all Asia since 1939, and have never relinquished absolute control of the Asians, through terrorisms, genocidal murders, rapes, robberies, and doping of survivors to become docile slaves laboring for the Japs and the Emperor of Japan.
Go to my website, dinosaurtoman.info, for more.
I know your reasons for doubting: How could I know? You've got "loose screws in your noggin"
I admit laxity in doing class home works since high school, but I've spent my high-school days reading novels and watching movies about western life, besides listening to lectures of teachers at school every class day. By night after school however, Salvador Formalejo, under pretence of friendship and enjoying my company, kept me in downtown Naga, away from home and rest-sleep till early dawn…resulted to my emaciated body and addiction to nicotine through the unending supply of cigarettes from Salvador, after only a few years. Salvador was most likely advised by other gangmembers on corrupting particularly me, and is still an active gang member; used by this gang of drugpushers, robbers, rapists…to destroy me and younger siblings since childhood, due to Salvador's close associations with our elder siblings in the past. I was already a 5-year full-blown nicotine addict without my knowing, when I went to Manila to begin

that always consisted of Salvador Formalejo, Florencio Celino and Sirit Celino, Manuel Rosales, Manuel Marpuri, Jesus Caletisin and Tirso Brocales, Balbino Mendes and son Jesus, Juanito Briones, Rodrigo Bigay, Jaime Andayog, T. Solo...doped Dr. Jun every night, at Manuel Marpuri's den after a whole day of doctoring service, till early dawn - with Dr. Jun semi-conscious with dope overdoze that Dr. Jun was carried home by Jesus Mendes and Juanito Briones...every time.
After a year of work in Manila, I returned to Ocampo intending to assist our parents reorganize and put some sense in ways of handling our home finances, and assisting in our younger siblings' studies. I did what I could amidst the vehement oppositions of our dope-addict parents whose thinking brains were already disabled by dope; was also not paid much attention by our younger siblings due to these gang had already turned our siblings' minds against me, after working and influencing our siblings since their earliest childhoods. And (Sister) Cena, on her rare visits to Ocampo from Surigao del Norte (with Leonora (Quobic) Edera and Cena's adopted daughters (Quobic's 2 nieces, who were sent by Manay Cena through college)) also vehemently opposed my intervening in the Dr. Jun's associations with this gang every night, justifying Dr. Jun's behavior as merely "unwinding on beer" (that was usually spiked by this gang), with "cousins", boyhood "friends", etc. – making me believe that even my sister, Cena (a Judge in Surigao del Norte) was also doped by Quobic and Quobic's family who, I've known are also Japs and Jap minions with metal-scrap business in Manila. And Quobic even addressed Fustino Marpuri as "Manoy (Brother) Tino".
Due to a cracking sound in my chest whenever I straightened up, I thought I had TB after college…so after some research, I've learned ODINAH tablets could cure the TB. I bought a 100-tablet bottle of the Odinah in a Naga pharmacy and began to take in the tablets according to the instructed dosage on the bottle-label. But near the depletion of the 100 tablets, I must have doubled the dosage after forgetting that I had already taken the tablets for the day…resulted to "drug over doze". And these dopepushers-liars convinced our parents and siblings that I've gone "coconuts in the noggin" (to their advantage, of course), due to the fact that I was and still am the only member of our home who remained inaccessible to their doping and lies,

since 1962 to this day! I could recall very little fragments of my experiences during the Odinah- over doze period. What kept nagging me to do something about these dopepushers-robbers-rapists who victimized our whole family continually since 1945 to this day, is that not one of them are suspected or apprehended or even "caught" by lawmen (due to local policemen were and still are "appointees" and close relatives to the gang leaders and members themselves, who had been voted by these gangmembers, to continuously hold Mayor or Vice-Mayor positions in the town hall). And these criminals have succeeded to destroy our family through continual doping, robbing, raping, lies, and enslavement of Dr. Jun and Engineer Achil in Ocampo since 1973, Hon. Judge Cena in Surigao del Norte since 1968, our parents since 1945 till both passed away in about 2000…

At first, The Bombase-Marpuri families led this gang during Mayor Antero Flor's term in the 1940s. Imelda Bombase was Vice Mayor and with help from Imelda's brothers – Juan Jr. and Domingo (whose in-laws are the bums and Naga hoodlums Marpuris – Bernardo, Gaspar, Totoy and relatives – Gervacio & sons, Fustino & brothers…), as well as the Bombase bum-squatter neighbors, Jesus Caletisin, Ricardo Martines, Balbino Mendes; and the Bombase relatives Oscar Asetre and Agaton Rosales; Imelda succeeded to dope Mayor Flor into a silly and disabled mayor thus leaving the town management to the Bombase-Marpuri crooks and hoodlums – I could recall Mayor Flor "high-in-dope", late at night promenading, singing noisily with Imelda, passing our old home, and stopping at Oscar Asetre's house (another bum in-law relative of the Bombases, Oscar's wife is a close relative of Agaton Rosales), in San Isidro Street. This gang continued successfully doping Antero Flor even after Flor's mayoral term ended, with cover up – "drunkenness-with-alcoholic drinks" until Flor died in the 1960s. I've never known where these Marpuris came from, their former victims if any, except that Domingo Bombase's wife is Leticia Marpuri who lived with the Bombases with her brothers Bernardo (a Naga home-made gun seller, possibly dopedealer too, and hoodlum), Gaspar, and an old widowed mother)

With these first experiences in dopepushing, this gang turned next on our parents and home, as their next victim. Ricardo Martines who was courting Erlina – Biyay and Bentong's daughter, used Biyay to dope Mom during childbirthing of our little sister, Lourdes (who

died a few weeks or months later due to abandonment) in 1945 after Dad accepted Biyay's offer as house-help to care for Mom. Biyay then robbed our home of undetermined amount of cash, jewelries, and titles to lands and cattle, (accumulated by our parents since their marriage in 1934, as well as by Mom's parents, due to Mom had lived and cared for her old father till he passed away, being the youngest child) that these gang later grabbed and rustled, plus the twice-yearly harvests on our about 33-hectare mostly ricelands, from 1945 till our parents passed away in the 2000s – almost 60 years of continual doping and robbing of our home.
Afterward, Ricardo Martines and Ambrosio Pellas (brother-in-law of Jesus Caletisin) were appointed policemen by the Gervacio Marpuri who became town mayor. Jesus Caletisin, Jorge and Florencio Celino, Domingo Bombase and Edmundo Asetre, the Bombase sisters, Imelda...were able to rent apartments in Naga and enrolled in UNC and Ateneo de Naga, without any source of income other than doping and robbing Mom and Dad, until all finished college, except the bum-rapist Jesus Caletisin who remained a bum even at school. Oscar Asetre, our old home neighbor, also was able to buy a complete ricemill with Diesel engine and sent daughter, Mely to finish Education at UNC, as well as support his large family of 12 without any other source of income except Paring's seamstress job, and doping and robbing of our home. Dad was often invited by Oscar to play "cards" in Oscar's house – Dad always came out doped though all of us in the family were made to believe that Dad was just "drunk" again.
Ricardo Martines' bum-squatter family consisting of brothers, sister, and old parents suddenly bought electric-generator and sound-system service to town dances and celebrations; suddenly built a 2-story home in Biyay's squatter lot by the river in Iraya (north of town), and also built a house where they squatted in San Isidro Street, made of materials from our old house that was dismantled after Dad learned of the lose of our home savings and family jewels that included Dad's gold ring with a large blue semi-precious stone... The Marpuris also built large houses from the loot stolen from our home. Our new home at San Isidro Street was never built, although concrete foundations were already set, 14 large trusses were already stocked on our "media agua", but the posts and lumber for walls and floors were never delivered by Gervacio Marpuri and sons, although Dad had paid-in-advance, thousands of pesos

to Gervacio Marpuri, who became the town mayor later, voted to office by these gang of town bums, robbers, dopepushers, rapists...and Dad has already lost his clear senses and thinking to dope. Dolfo Belaos was the buyer of the rustled cattle (22 carabaos, goats...) and could also deal in dope for the gang, due to Dolfo's Manila connections. And Dolfo, Bernardo Marpuri were the operators of a cockfight arena in town, where all these bums, drugpushers and their doped victims (well covered-up with "drunk" of alcoholic drinks), assembled every Sunday.

Also every Sunday, the elder members of the gang met at Renato Villamora's spacious ground floor, for a church "cursillo" assembly. Dad always came out of these meetings doped ("drunk"). I believe Renato Villamora (a son married Marino's sister) was killed by Marino Marpuri, Bernardo Marpuri, Domingo Bombase, Jesus Caletisin...members of the "cursillo", after this gang openly displayed spiking Dad's drinks with dope to Renato V; being an honest devout churchgoer (a son is an ordained priest) must have reported Dad's doping to Rev. Jose Rey, so Renato V (in his early 60s) became a dead Renato – for being a danger to exposing the criminal activities of the gang. The same happened to Alfredo Cortes, who got into Manuel's den to play "cards". Alfredo was never allowed to leave the den alive. This gang must have overpowered him and forced him to drink heavily spiked beer, resulted to his semi-conscious and partially paralyzed physical state. Or these gang members (Rodrigo Bigay, Balbino Mendes, Jesus Caletisin, Francisco Marpuri, and other elder brawny gang members like Bernardo Marpuri, who had experiences in disabling victims must have managed to break Alfredo's spine in several places thus accounting for Alfredo's semi-paralyzed state that left him seated in the den, for days. He remained seated on a chair in the den until he expired. No one among his in-laws dared to get into the den to aid Alfredo due to all members of his family knew and were terrorized by this gang. Vice Mayor Francisco Marpuri who lived beside Manuel's den must have known everything happening in the den, including Dr. Jun's doping...

Imelda Bombase received the stolen jewelries from our home and other victims by the gang, due to Imelda was able to open a jewelry store beside Alex Theater in Naga City in the 1950s as well as rent an apartment near Collegio de Sta. Isabel where Edmundo Asetre and some boarders also lived at the first few months but left

the house because Imelda entertained male guests nightly, with drinks. The boarders were Raul Villamora, the Lim brothers of Hanawan, the Pasibe children of Goa who left immediately after a few weeks.
Enso and Charing Formalejo also never paid back a 300-peso loan from Mom and Dad, the 300-peso loan spent on buying a rice mill and a Ruston Diesel Engine in 1944. Enso who was a Jap minion, and Charing whose parents were migrants from Ilocos during President Quirino's term, just wrote off the loan after Mom and Dad were doped. Enso's brother in-law, the Jap Agaton Rosales and two brothers were chasing off the other tenants in Pedro Briones' 148-hectare lands, in preparation for the implementation of Jap Dictator Marcos' Land Reform Program. All the Japs and Jap minions in the Philippines were tipped off by Marcos, about when the Land Reform will take effect to give the tenants land titles to the lands they till – free of charge from the former landowners. At the same time that Dictator Marcos and wife Imelda were confiscating lands from Spanish Grandees (one was Don Mariano's 7000 hectare lands in Partido that was renamed "Hacienda Magdalena") in the Philippines, depositing the titles in the Philippine Treasury where the wife, Imelda Marcos emptied the Treasury and shipped Cash (including US $AID of $400 Million yearly since 1945…) by the planeload, to the Marcos' Swiss bank accounts, several times a day for more than a decade. And the hundreds of thousands of hectares of confiscated lands were turned over to the Marcos Ill-Gotten Holdings, to the loud vehement protests of the Marcos' cronies who finally revolted and walked the "Pennsylvania Avenue" to the Marcos Palace in Manila, waving fully-loaded Tommy Guns, crying out "Share the loot! Share the loot," then the US Cavalry aboard an unmarked black chopper from the JFK Aircraft Carrier in Manila Bay, lifted off Marcos and Imelda from the palace roof with the Marcos Ill-Gotten holdings and Swiss-Bank account deposit slips, inside dozens of large, bursting to the seams, suitcases just as the guns started blazing at the palace gates, thus saving the Marcos family from getting riddled with lead from their disgruntled former cronies in the 16-year dictatorship.
And all the Japs and Jap minions in the Philippines who got free land titles to the lands they tilled, become the present small and big landowners in the Philippines, including Agaton's 50-hectare irrigated ricelands, and Agaton's sister's (who married Pedro Briones' Overseer

Juan Iraula) 39-hectare ricelands - both ricelands, a part of Pedro Briones' 148-hectare lands in Ilawod, Ocampo. And the Marcos Holdings remain intact in the Philippines after a Marcos daughter married into the Jap Aranetas, (Feati U, Hydro-Electric Power Plants, Boilers, Colgate-Palmolive, Del Monte, Mobil...) thus successfully "money laundering" Marcos ill-gotten wealth that included Manila Electric which was taken over by Marcos from the Lopes families... And my Mom, Dad, and all my siblings got all messed up with long-term doping, robbing, and enslavement by the Japs, Jap minions, and Marcos province-mates who were dispersed all over the country as migrant bums and criminals from the 1940s till today. Most of the drugpushers, bums, robbers etc. in my old hometown were migrants from Ilocos (where Marcos came from). And the Filipino former landowners and victims of doping, robberies, rapes... were helpless due to Marcos disarmed all the "non-Jap natives" then created the Philippine National Police that consisted of the bums, Jap minions, criminals from Ilocos...copycat the Chinaman (Japs disaguised as Chinaman) Police in China, that consisted of the Jap military in China and of the Jap minions among the Chinaman. And the Marcos loots, the Jap landowners through Marcos Land Reform, Imelda's Sorority, and Marcos' hatchet men, the Philippine National Police remained INTACT in the Philippines since the 1960s to 2010 AD!
CRIME PAYS!!
And Marcos himself was a convicted murderer shooting a father's rival for governor of Ilocos, and served time in prison for the crime, before becoming President.
Years before the assassination of Marcos' rival to the presidency – Benigno Aquino who went to teach in Harvard USA, Marcos threatened that Aquino could never step again on Philippine soil, alive! When Aquino returned and stepped out Pan Am at the Manila International Airport, a policeman shot and killed Aquino at the PanAm's threshold, then the shooter was also shot and killed by another policeman in a double-cross, resulting to "dead-end" investigation (Case Solved.) made by Marcos investigators!
The local criminals in Ocampo were just small-time criminals; component of big-time criminals of the Marcos family, the Jap criminals in China, Koreas, Mongolia...– Emperor of Japan and all Asia! – The Biggest Big-Time Criminal on the planet! Saddam Hussein was a baby compared to this Emperor.

Where did I find all these information? – over heard private conversations between Mom and Dad; from conversations with friendly oldsters in Ocampo (examples – Pedro Briones' concubine Awing – Alfredo Briones 70-year old mother…other landowners in Ocampo), from newspapers-magazines in Naga and Manila libraries, and from my associations with gangmembers from 7 to 18 years old; also from information I've gleaned and remembered in the events around me, since my childhood to my adulthood in Ocampo, Manila, and in the U.S.A.

And I, my siblings, and parents who NEVER have knowledge of dopes, dope addiction, dopepushers…became easy victims by these criminals for about 60 years from 1945 to this day. I, alone, regained my lucid thinking and was able to know about these dope pushers, when I absolutely severed contact with these criminals, cousins, "friends," from Ocampo since 1962 to the present. And I also learned about illegal drugs, drug addictions, alcoholism… in 1997 during training as Correctional Officer in Lake Butler FL. Hence, my recollections of my parents behaviors, the behaviors of these drugpushers-robbers, other personal experiences with Salvador Formalejo and other gangmembers, events in the old hometown and in Naga, in Manila…all add up for me to come to a conclusion: that our family was the victim by these criminal gang since 1945; that these gang used illegal drugs on our parents regularly, in order to rob, rape, abuse our parents; that our parent's senseless behaviors since 1945 until both passed away in about 2000 were due to the effects of the drugs in their systems; and that our elder siblings and myself, since the 1950s, were also doped without our knowing, and this gang also doped our younger brothers Dr. Pastor Jr. after graduation in 1972 in Manila and since 1973 to 1981 in my sight, during the doctor's medical service in Ocampo, until the doctor left for the U.S.A. in 1990–a destitute invalid, ruined professionally and physically due to long-term doping and robbing by this gang.

And this gang eluded suspicion and apprehension because the usual cover-up of "drunk with alcoholic drink" was very effective and believed by our innocent family members; and the town-hall officials and town policemen were all members of this gang, since 1945 till today.

For instance:

Imelda Bombase (Vice Mayor 1944) and Domingo Bombase held government positions in the town hall, and appointed gang members as armed policemen (Ricardo Martines, Ambrosio Pellas…? Umali) after the members initially doped, robbed, and raped their victims; Gervacio Marpuri (Mayor 1950s) appointed policemen (? Tawagon), Francisco Marpuri (Vice Mayor 1970s), Rolando Belaos (Mayor 1970s) appointed policemen (Carlos Ibarientos, - Andayog, Pedro Ocampo) all these policemen are gangmembers, Japs, and Jap minions, and the only armed "Filipinos' in town. I've heard in the USA that Florencio Celino – another long-time gangmember who was among the dopepushers who victimized Dr. Pastor Jr. since 1973, became town mayor after I left the old hometown.
Some local innocents attempted to give helping hand to me, to Dr. Pastor Jr. as true friends, but all ended up murdered (Teofilo Yamson ?, Renato Villamora, Ricardo Pangilinan, Jose Baliuag, Alfredo Cortes…) or "bought" and/or transferred to other towns (Wilfredo Briones, Rev. Jose Rey…).
So this gang of criminals have their criminal activities successfully ongoing since 1945 to this day!
THE VICTIMS

1. Antero Flor (Mayor 1944…)
2. Francisca Alcantara & Husband Federico Pena (1944…)
3. Fidela Alcantara & Pastor Briones Sr. (1945…2000)
4. Pastor Briones Sr' Children–Dr. Pastor Jr. 1972 Manila…1973 Ocampo Engineer Achil 1974 Manila…1976 Ocampo
5. Eustaquia and Talina Borromeo (1950s)
6. Loiue Briones (1950s)
7. Alejandro (Enso's relative)–wife, son, and tenanted land-all stolen
8. Unknown others

FAMILY TREES
BRIONES
Leocadio–Captain, Spanish Civil Guard, 1700s
"Borad"–Leticia Nabata
Filomeno migrated to Irosin, Sorsogon, 1945
Agosto-Araceli (daughter Estrella)
Teofilo
Teoy (Pastor's brother) died under cattle hoofs as a child
Pastor–Fidela Alcantara
Azucena–Judge, Surigao del Norte
Elias–Estela M–USN Retired, (Las Vegas, NV USA)

Children (Daryl, Denny, Karen Allen...)

Nic (TX USA)

Lourdes–died as a baby due to abandonment

Pastor Jr.-Doctor of Medicine UST '73 (Gang Victim)

Damaso–Thelma P, Mandaluyong Philippines

Children (Gail, son1, son2, son3)

Teofilo–Reggie S (LA, CA USA)

Children (Ryan, Sean)

Yolanda–Renato V (Gainesville, FL USA)

Child (Bianca Rhea)

Achil, Mechanical Engineer UST'75 (Gang Victim)

- Sandra, Children (Girl 1, sons...)

Dad was orphaned at 16; worked as Detention Officer, Naga City Jail;
Was "Encargado" of Don Mariano Garchitorena's lands; then was "Capataz" of road building in "Rinconada" until the 1950s. Dad never knew that Mom was doped in 1945 and never suspected that Mom's sudden changes in behaviors from a loving housewife to unexplained absences from home, after 1945, were due to Mom's being victimized by the members of this gang. And Dad never knew also that this gang had doped and robbed...our home during Mom's childbirthing of our little sister, Lourdes who died a few months later due to abandonment.
Mom never worked to earn a living before 1945, due to her inheritance from her parents, was more than enough to meet her needs. Like Dad, Mom never smokes, drinks alcoholic beverages. And our siblings were never "drunks" or alcoholics since our childhood days – only after these gang members doped Dr. Jun, Engineer Achil...(Dad) by heavily spiked beer and drinks continually for years...they had lost their usual brilliant thinking brains (as students).
Both Mom and Dad were honest, guileless, and simple folks enjoying their works and lives before 1945. This gang however, succeeded to convince me (till the late 1980s) and my siblings that Dad and Mom were irresponsible, negligent parents...

Andres-?

Luciano

Pedro-?

Sons… Jacobo-Naty O
Sabino (present Heir 250-hectare land) 60s

ALCANTARA
Mariano–Letty Cortes
Sons, Daughters-Built and founded the town, Ocampo in the late 1800s; each of the children were given 24-hectare lots. Fidela, the youngest lived and took care of aged Mariano until his death in 1934.
Francisca–Federico P
Purita–Leoncio E 80s
Sons…
Loreta–Fustino Marpuri (Bum) 80s
Sons, Daughters
Manuel–Betty A
Sons, daughters
Fidela–Pastor B
Sons, daughters
Charing, Pina, Itas 90s (sisters)
Salvador, Salvacion, Rosita
Agapito, Manuel, Michael….
Benigna, Ditas, Lourdes
Demetrio, Perdon 90s
Children
Salvacion, Modesta 80s (sisters)
Children
Aurelia–Losa 90s
Children
Fustino and his BUM family took over his wife's (Loreta) family, occupied and lived in Loreta's houses, managed to overdoze with dope Loreta's father into a drooling laughing invalid idiot, and began to rob his in-laws of twice-harvests from ricelands…to support and give capital to his brothers, sisters, and mother; then turned on robbing my Mom and Dad through his brother's in-laws (Andong and Sela Botor).
GANG
Bombase–Marpuri
Domingo–Leticia M (Brothers: Bernardo, Gaspar…all bums 70s) 90s
Juan Jr (Bum) 80s
Imelda (widow) (Vice Mayor 1944) 80s
sons 2
Siang–Andong Asetre (Oscar's brother, bum) 90s
Edmundo (bum)
Daughters 2
Asetre–"Rosales" 80s

Melicio (bum)-Seduced a Beltran senorita from Caramoan.

Elsie–married to Ibarientos (tenants in Briones' lands)

Sons, 8 daughters 60s

Marpuri

Faustino & brothers, sisters, mother (all Bums) 80s

Gervacio Town Mayor 1950s

Marino, Francisco, Pepe...80s

Sons & daughters (migrant bums)

Caletisin (BUMS-SQUATTERS)

Jesus–rapist, cardsharp cheater 80s

Salvador

Ambrosio P. in-law (police, Ocampo)

Salvador S. in-law (tenant Jap Mendosa)

Tirso Brocales (bum nephew of Jesus) 60s

Martines (BUMS-SQUATTERS)

Ricardo (police, Ocampo) 80s

Parents, brothers, sister

Mendes (BUMS-SQUATTERS)

Balbino–Talina M 80s

Sons, Jesus...60s

Celino (BUMS-SQUATTERS)

Jorge, Bernardo, Sirit 60s (Shown pounds of white powder in plastic bag, several times during Dr. Jun's doping in Manuel Marpuri's den...

Florencio (Town Mayor 1980s) 70s

Rosales

Agaton-Pina 80s

Agapito 60s

Manual-Belaos

Sons, daughters

Formalejo

Enso-Charing 80s

Salvador (Bum)

Salvacion (Practicing doctor in Manila)

Rosita

Belaos 70s

Rolando-Jallores (Town Mayor 1970s)

Rodolfo (Rustler...)

Children

Rolando Belaos was town mayor when this gang began doping Dr. Pastor Jr. every night since 1973 (Francisco Marpuri was Vice Mayor). And Rodolfo Belaos, his uncle was among the active participants in the nightly

doping and robbing of Dr. Pastor Jr's daily earnings of about 250 pesos.
Andong Asetre died in the 1970s-a dragon-hallucinating wreck after getting doped by Domingo-Leticia Marpuri and brothers, due to the bum Andong never liked to work to support his family, and was just another mouth fed in the large Bombase-Marpuri household.
These policemen were appointed and armed by the gang leaders who were voted to become town-hall mayors…by the bum migrants in town, after doping and robbing victims in town.
Other gang members: All about 60 years old by 2010

Rodrigo Bigay (ex-convict?)
Juanito "Briones"
Jaime Andayog
T. Solo
Romeo Patulot
? Rustia
? Cesar
Severo Polo (police, Queson City)
Kulas (Severo's uncle 80s)
Carlos Ibarientos (police, Ocampo)
? Andayog (police, Ocampo)
? Tawagon (police, Ocampo)
Willie Trinidad

Tenants and squatters in Pedro B's lands in Ilawod: Olos, Iraula…
All these gang members and leaders are Japs, Jap minions and migrants from Ilocandia where the dictator Marcos came from.
The Bombases lived in one house with Domingo's and Juan Jrs' wives and children, with Imelda and her two sons, with Edmundo Asetre and his parents, with the brother of Domingo's wife: Bernardo, Gaspar Marpuri, and with the aged parents and with two unmarried sisters of Domingo. And their only incomes were Domingo's town hall job and Imelda's Vice Mayor work, and Domingo's wife – Leticia Marpuri's teacher's salary. But after 1945, Domingo, Edmundo, Bernardo, the Bombase sisters, and Imelda's two sons were all going to schools in UNC and "Ateneo de Naga" and renting apartments too, in Naga.
Jesus Caletisin, Jorge Celino, and Florencio Celino (all bums also with no work and incomes), also studied in UNC and rented apartments in Naga, since the late1940s till 1958 where I saw them living in an apartment beside the old Bragais Studio. These dope pushers, robbers and rapists supported themselves in

the continual victimizing of Mom and our home since 1945.
The Bombases-Marpuris also robbed Dad of cash-proceeds from Marcos' Asian Development Bank loan to build Dad's cone-type rice milling business, whenever Dad withdrew the loan money from the bank in Naga– Dad came several times to where I and Jun lived in Naga, heavily doped that I, at that time believed as only "drunk" with alcoholic drinks (the usual "cover-up" that these gang members spread about to my siblings, who still believe so, even to this day).
Florencio Celino, who was always with this gang in doping and robbing Dr. Jun in Ocampo, wrote me a letter after getting elected as town mayor in late 1989, asking me to provide Cash for the "betterment of the town", as if he was not among the dopepusher-robber I've known since his bum-squatter days until he seduced a she-devil for his wife in 1987.
Jorge Celino succeeded to get a job in UNC as Music Band leader, that Jorge passed down to Florencio after Jorge stopped schooling to marry an old spinster with money (Flora Tenorio); and Florencio became the only support of his mother and three younger sisters who were also studying at UNC without any other family income except Florencio's UNC job and his share in this gangs' criminal activities in the town.
I believe that with the first robbing of Dad's rice lands in San Vicente by Ricardo Martines, the twice yearly harvests from the rice lands were shared by all the members of this gang to take care of their otherwise starving families, since 1945 to this day!
65 years of crimes of parents passed on to children, nephews, nieces…and down to grandchildren! If such fiendish actions are normal, then what's abnormal? In my humble opinion, these recidivist gang members, criminals must be put apart from man's society into mental institutions for the CRIMINALLY INSANE!
HAPPIER DAYS
1944: Mom would rock me in a wooden cradle, singing me to sleep with her sweet mellow voice: "Fairie Moon shines again on the prairies, And the cowboys are riding today" Repeated over and over again.
Blindfolded with handkerchief, turn around a few times, then instructed to walk to a selected known spot nearby.
Dad's HealthCare (1944): spoonful of Liver Oil and Weightlifting of 2 train wheels still attached to axle, lifted several times in weekends.

Riddles given by Dad, during playtime on a sleeping mat, before we sleep:
"In the middle of the sea, there's a yellow taxi" Egg Yolk.
" Mississipi is a long word, but can you spell "it"?" I T
Our dining room wall had two beautiful framed photos of a fully cooked fresh-water giant crab all red in color, with open pincers and legs, all set on a blue and green decorated bone-china platter. At one end corner, was a large green and yellow talking parrot inside a cage. The parrot got excited when a visitor approached our home, and cried out shrilly "kakagit! kakagit…" ("Cuidado! Beware…"). The parrot, the photos, the cradle, the toys etc. disappeared after the doping of Mom in 1945; along with the trusses stocked on our "media agua" some of which reappeared in Biyay-Bentong-Ricardo's newly built house in the north (iraya) of town).
TOYS
A rabbit in box; press a button and the rabbit pops out, laughing.
A uniformed drummer boy stands on cart. Pull the cart and the boy beats the drum.
Clay roosters and hens painted real-life with glossy multi-colors.
Home Larder: Always filled with food – Canned ChopSuey, Carnation Milk, Quaker Oats, Fresh aromatic Apples wrapped in thin papers, grapes in the vine…all disappeared and never reappeared after 1945.

Conclusions

The Evolution of a life form i.e. from trilobite to fish or from a biped dinosaur to biped Man, is due to adaptations of the life form to better survive in changes of environments. The life form developed different traits in features, physical attributes and brain capabilities to get more food provided by the environments.
The breakup of the supercontinent, Pangea, 220 million years ago has changed the oceans and landforms on planet Earth...This is the story of the Dinosaurs and the Mammals that thrived on the ongoing changes of the landforms and the environments of planet Earth.